Scalable Data Analytics with Azure Data Explorer

Modern ways to query, analyze, and perform
real-time data analysis on large volumes of data

Jason Myerscough

BIRMINGHAM—MUMBAI

Scalable Data Analytics with Azure Data Explorer

Publishing Product Manager: Sunith Shetty
Senior Editor: Roshan Kumar
Content Development Editor: Shreya Moharir
Technical Editor: Sonam Pandey
Copy Editor: Safis Editing
Project Coordinator: Aparna Ravikumar Nair
Proofreader: Safis Editing
Indexer: Manju Arasan
Production Designer: Aparna Bhagat
Marketing Coordinator: Priyanka Mhatre

First published: March 2022

Production reference: 2150322

Published by Packt Publishing Ltd.
Livery Place
35 Livery Street
Birmingham
B3 2PB, UK.

ISBN 978-1-80107-854-2

www.packt.com

To Harrison and James, my two boys. I love you more than you will ever know. The stories and comics you both write inspired me to write this book.

To my parents, Charlotte and Dave. Thank you for getting Damian and me a PC and letting us watch Hackers. That was the catalyst for everything.

To my mentors, John Vasicek and Arunee Singhchawla. Thank you, John, for starting my Azure journey, supporting the original DevOps movement, and trusting me to lead it. Thank you Arunee, for your mentorship, for encouraging me to write this book, and last but not least, friendship. I learn from you every day.

Thank you, Shreya Moharir, for all your help and feedback. I learned so much from you during this project.

Foreword

It has been almost three decades since I started my career in technology. It started from the fact that I am a curious person and always want to find various ways to solve problems. I love the challenge of finding new ways to do things more efficiently. By doing so, I tend to use the information around me to help decide on the approach that would be best in solving problems. I entered the IT world as a quality assurance analyst, and then that flourished to site reliability engineering. What I have learned over the years is that those positions share a common property, which is analysis. My passion is in using data and information at hand to aid in the decision-making process. My passion for analysis continues to thrive. Throughout those years, I have been blessed to have met and learned from many creditable people.

In 2016, I had the privilege of meeting Jason Myerscough. He joined the team with the purpose of transforming and modernizing the team. Jason pioneered the DevOps culture and was tasked with leading one of the biggest and most notable projects: migrating one of the flagship products to the Azure Cloud. I had the honor of joining his team to help form this initiative. I remember how excited and nervous I was to be part of such an initiative for our team and company. I remember watching him sift through all the data and analyzing it. I recall him walking across the hall back and forth looking at the data and discussing with our data scientists how to formulate the pattern from the large amount of data he collected from production. He then would come up with a monitoring and alerting strategy, as well as performance tuning. I wish I could be like him, I thought to myself. With his strong development background combined with his curiosity, strive for excellence, passion for the Azure Cloud, and analysis skills, I knew we had the perfect person who would bring us to the cloud.

These days, we live in a data-rich age. It has become much more critical to be able to extract and digest data in a meaningful way in order to make business-critical decisions. As a site reliability engineer, some of the key responsibilities are analyzing logs, monitoring production environments, and responding to any issues. It is even more critical for us to be able to have in-depth insight into production data, to be able to digest and set up proactive alerting.

In Jason's book, he shares how you can use Azure Data Explorer to quickly identify patterns, anomalies, and trends. His book walks you through what Azure Data Explorer is, how to set it up, and how to use Kusto Query Language to run as many queries as you need to quickly answer your questions.

With Jason's in-depth experience as a developer, as a site reliability engineer, and as an architect, this book truly reflects his experience and passion. It captures perfectly how you can use Azure Data Explorer to analyze data-rich environments and have a meaningful way to help make the right business decisions. I highly recommend this book as I find it full of various important aspects of Azure Data Explorer. It is such a delightful and easy-to-follow read. If you have a passion for analysis or are in a position where you must make business decisions based on large data volumes, then you should put this book on your required reading list. This book will empower you and your team to improve efficiency and productivity.

– *Arunee Singhchawla*

Director of Site Reliability Engineering at Nuance Communications

Contributors

About the author

Jason Myerscough is a director of Site Reliability Engineering and cloud architect at Nuance Communications. He has been working with Azure daily since 2015. He has migrated his company's flagship product to Azure and designed the environments to be secure and scalable across 16 different Azure regions by applying cloud best practices and governance. He is currently certified as an Azure Administrator (AZ-103) and an Azure DevOps Expert (AZ-400). He holds a first-class bachelor's degree with honors in software engineering and a first-class master's degree in computing.

A special thanks to all the Microsoft team members—Vladik Branevich, Tzvia Gitlin Troyna, Adi Eldar, Pankaj Suri, Dany Hoter, Oren Hasbani, Guy Reginiano, Slavik Neimer, Michal Bar, Rony Liderman, and Gabi Lehner for contributing to this book with your valuable insights.

About the reviewers

Diana Widjaja has been a technical writer with Salesforce for over 10 years, where she writes developer documentation for teams across the company's UI platform. Previously, she also had the opportunity to work with CBS, IBM, and Google. Diana received a bachelor of science in technical communication from the University of Washington and a master of science in information security policy and management from Carnegie Mellon University. When she's not working with words, pixels, and code, Diana likes to explore the San Francisco Bay Area with her energetic kids and husband, and keep in touch with family in England and Singapore.

Sibelius dos Santos Segala is a software engineer. He has been involved in IT projects for over 25 years in diverse areas such as university logistics, HR and grade systems, mainframe application integration to intranet, multimedia application streaming in consumer devices, support application for automotive tracking, and lately, banking-related software.

Table of Contents

3
Exploring the Azure Data Explorer UI

Section 2: Querying and Visualizing Your Data

4
Ingesting Data in Azure Data Explorer

5
Introducing the Kusto Query Language

6
Introducing Time Series Analysis

7
Identifying Patterns, Anomalies, and Trends in your Data

8
Data Visualization with Azure Data Explorer and Power BI

Section 3: Advanced Azure Data Explorer Topics

9

Monitoring and Troubleshooting Azure Data Explorer

10

Azure Data Explorer Security

11
Performance Tuning in Azure Data Explorer

12
Cost Management in Azure Data Explorer

13
Assessment

Index

Other Books You May Enjoy

Preface

Azure Data Explorer (**ADX**) enables developers and data scientists to make data-driven business decisions. This book will help you rapidly get insights from your applications by querying data at scale and implementing best practices for securing your ADX clusters.

The book begins by introducing ADX and discussing its architecture, core features, and benefits. You'll learn how to securely deploy ADX instances and be comfortable navigating and using the ADX Web UI. You'll focus on data ingestion and how to query and visualize your data using the powerful **Kusto Query Language** (**KQL**). You'll cover KQL operators and functions to efficiently query and explore your data. You'll learn to perform time series analysis and how to search for anomalies and trends in your data. Later, you'll focus on advanced ADX topics, starting with deploying your ADX instances using **Infrastructure as Code** (**IaC**). You will manage your cluster performance and monthly ADX costs by handling cluster scaling and data retention periods. Finally, you will cover how to secure your ADX environment by restricting access using subnet delegation and cover some of the best practices for improving your KQL query performance.

By the end of this book, you will be able to securely deploy your own ADX instance, ingest data from multiple sources, rapidly query your data, and produce reports with KQL and **Power BI**.

Who this book is for

This book is for data analysts, data engineers, and data scientists who are responsible for analyzing and querying their team's large volumes of data on Azure. This book will also be helpful for SRE and DevOps engineers that are responsible for deploying, maintaining, and securing the infrastructure. Some previous Azure experience and basic data querying knowledge will be beneficial.

What this book covers

Chapter 1, Introduction to Azure Data Explorer, covers what ADX is, the core features of ADX, and where ADX fits in Microsoft's suite of data services. The chapter then discusses some of the different use cases of when to use ADX and demonstrates how to execute your first KQL query.

Chapter 2, Building Your Azure Data Explorer Environment, explains how to quickly deploy and configure ADX clusters and databases using the Azure portal, PowerShell, and Azure ARM templates. By the end of this chapter, you will be ready to start ingesting and analyzing your data.

Chapter 3, Exploring Azure Data Explorer UI, presents the ADX UI to you. You will spend the majority of your time using the ADX UI to query and analyze your data. By the end of this chapter, you will be familiar with the windows and panes in the ADX Web UI.

Chapter 4, Ingesting Data in Azure Data Explorer, discusses the concept of data ingestion and demonstrates how to ingest data from multiple data sources such as Blob storage and Azure Event Hubs, how to create new table schemas, and explains how data maps to those tables. At the end of this chapter, you will understand how ADX ingests data and how to configure the data ingestion.

Chapter 5, Introducing the Kusto Query Language, introduces you to KQL and demonstrates how to query data. The chapter begins by introducing the language, explains the basics of KQL such as searching, filtering, aggregating, and joining tables. By the end of the chapter, you will know enough KQL to comfortably query data.

Chapter 6, Introducing Time Series Analysis, introduces you to ADX's time series features, beginning by defining what time series analysis is, and then demonstrating how to query your time series data using the make-series operator. Finally, we discuss some of the most important and useful time series functions provided by ADX.

Chapter 7, Identifying Patterns, Anomalies, and Trends in Your Data, builds on the previous chapter by discussing how to detect anomalies and trends in your data. The chapter first begins by introducing some of the anomaly functions available within ADX and then covers some of the machine learning capabilities of ADX.

Chapter 8, Data Visualization with Azure Data Explorer and Power BI, explains and demonstrates how to integrate ADX with Power BI. Power BI is a powerful reporting tool used to share rich graphs and reports. By the end of the chapter, you will know how to integrate ADX with Power BI and how to create reports in Power BI powered by ADX datasets.

Chapter 9, Monitoring and Troubleshooting Azure Data Explorer, teaches you how to monitor your ADX clusters using Azure Monitor and ADX Insights. The chapter teaches you how to configure alerts using KQL and action groups and explains how to troubleshoot issues by enabling the ADX diagnostics and examining those logs using Log Analytics. In the troubleshooting section, we will demonstrate how to troubleshoot and resolve a data ingestion problem.

Chapter 10, Azure Data Explorer Security, discusses how to secure your ADX instances using both identity management and virtual networks with subnet delegation. We begin by explaining why security is important on the public cloud and then we discuss identity management at the management and data plane. Next, we will introduce securing ADX instances using virtual networks and subnet delegation and demonstrate how to filter network traffic using **network security groups** (**NSGs**).

Chapter 11, Performance Tuning in Azure Data Explorer, begins by explaining why performance matters, and then discusses the KQL best practices and revisits the ADX architecture to explain how time filtering can provide performance improvements. You will also learn how to monitor the performance of your clusters, queries, and external applications.

Chapter 12, Cost Management in Azure Data Explorer, discusses how to plan and manage production deployments. The chapter first discusses how to manage your clusters and what requirements you should take into consideration when planning your deployment, and finally discusses how to estimate your Azure costs.

To get the most out of this book

To get the most out of the book, we recommend that you create an Azure account and take advantage of Microsoft's 30-day free trial to follow along with the practical examples. We will spend most of our time in the Azure portal, Azure Cloud Shell, and the Data Explorer Web UI. We also recommend that you clone the repository to your local machine and use Visual Studio Code to experiment and modify the code samples.

Software/hardware covered in the book	Operating system requirements
Visual Studio Code (`https://code.visualstudio.com/`)	Windows, macOS, or Linux
Git command-line tools (`https://git-scm.com/downloads`)	Windows, macOS, or Linux

If you are using the digital version of this book, we advise you to type the code yourself or access the code from the book's GitHub repository (a link is available in the next section). Doing so will help you avoid any potential errors related to the copying and pasting of code.

Please remember to turn off/deallocate your resources in Azure to avoid incurring extra charges.

Download the example code files

You can download the example code files for this book from GitHub at `https://github.com/PacktPublishing/Scalable-Data-Analytics-with-Azure-Data-Explorer`. If there's an update to the code, it will be updated in the GitHub repository.

We also have other code bundles from our rich catalog of books and videos available at `https://github.com/PacktPublishing/`. Check them out!

Code in Action

The Code in Action videos for this book can be viewed at `https://bit.ly/3uw1w2U`.

Download the color images

We also provide a PDF file that has color images of the screenshots and diagrams used in this book. You can download it here: `https://static.packt-cdn.com/downloads/9781801078542_ColorImages.pdf`.

Conventions used

There are a number of text conventions used throughout this book.

`Code in text`: Indicates code words in text, database table names, folder names, filenames, file extensions, pathnames, dummy URLs, user input, and Twitter handles. Here is an example: "Since we know there is a failed ingestion, the table we are interested in is aptly called `FailedIngestion`."

A block of code is set as follows:

```
StormEvents | where State =~ "California"
   | summarize event=count() by EventType | render columnchart
```

Any command-line input or output is written as follows:

```
Get-AzRoleDefinition | Select-Object Name, Description
```

Bold: Indicates a new term, an important word, or words that you see onscreen. For instance, words in menus or dialog boxes appear in **bold**. Here is an example: "Next, click **Review + create**. Finally, click **Create** once the validation is complete."

Tips and Important Notes
Appear like this.

Get in touch

Feedback from our readers is always welcome.

General feedback: If you have questions about any aspect of this book, email us at customercare@packtpub.com and mention the book title in the subject of your message.

Errata: Although we have taken every care to ensure the accuracy of our content, mistakes do happen. If you have found a mistake in this book, we would be grateful if you would report this to us. Please visit www.packtpub.com/support/errata and fill in the form.

Piracy: If you come across any illegal copies of our works in any form on the internet, we would be grateful if you would provide us with the location address or website name. Please contact us at copyright@packt.com with a link to the material.

If you are interested in becoming an author: If there is a topic that you have expertise in and you are interested in either writing or contributing to a book, please visit authors.packtpub.com.

Share Your Thoughts

Once you've read *Scalable Data Analytics with Azure Data Explorer*, we'd love to hear your thoughts! Scan the QR code below to go straight to the Amazon review page for this book and share your feedback.

https://packt.link/r/1-801-07854-8

Your review is important to us and the tech community and will help us make sure we're delivering excellent quality content.

Section 1: Introduction to Azure Data Explorer

This section introduces you to **Azure Data Explorer** (**ADX**) by discussing the core features and benefits of ADX, such as low-latency data ingestion, the ADX architecture, and how to quickly deploy your instance of ADX via the Azure portal, PowerShell, and ARM templates. The final chapter of this section presents an overview of the ADX web UI, where you will spend most of your time analyzing your data. By the end of this section, you will understand the core features of ADX, be able to deploy your own ADX instances, and be comfortable navigating and using the ADX web UI. This section sets the foundations for *Section 2*, where you will begin to ingest and analyze the data.

This section consists of the following chapters:

- *Chapter 1, Introducing Azure Data Explorer*
- *Chapter 2, Building Your Azure Data Explorer Environment*
- *Chapter 3, Exploring the Azure Data Explorer UI*

1
Introducing Azure Data Explorer

Welcome to *Scalable Data Analytics with Azure Data Explorer*! More than 90% of today's data is digital and most of that data is considered unstructured, such as text messages and other forms of free text. So how can we analyze all our data? The answer is data analytics and **Azure Data Explorer (ADX)**. Data analytics is a complex topic and **Microsoft Azure** provides a comprehensive selection of **data analytics services**, which can seem overwhelming when you are first starting your journey into data analytics.

In this chapter, we begin by introducing the data analytics pipeline and learning about each of the steps in the pipeline. These steps are required for taking raw data and producing reports and visuals as a result of your analysis, which will help you understand the workflow used by ADX.

Next, we will introduce some of the popular **Azure data services** and understand where they fit in the data analytics pipeline. Some of these services, such as **Azure Event Hubs**, will be used in later chapters when we learn about data ingestion.

We will also learn what **ADX** is, the features that make it a powerful data exploration platform, the architecture, and key components of ADX, such as the engine cluster, and understand some of the use cases for ADX, for example, in **IoT monitoring, telemetry, and log analysis**. Finally, we will get our feet wet and dive right into running your first **Kusto Query Language (KQL)** query using the Data Explorer UI.

In this chapter, we are going to cover the following main topics:

- Introducing the data analytics pipeline
- What is Azure Data Explorer?
- Azure Data Explorer use cases
- Running your first query

Technical requirements

If you do not already have an Azure account, head over to `https://azure.microsoft.com/en-us/free/search/` and sign up. **Microsoft** provides 12 months of popular free services and $200 credit, which is enough to cover the cost of our Azure Data Explorer journey with this book. Microsoft also provides a free to use cluster (`https://help.kusto.windows.net/`) that is already populated with data. We will use this free cluster and create our own clusters throughout this book.

Please remember to clone or download the Git repository that accompanies the book from `https://github.com/PacktPublishing/Scalable-Data-Analytics-with-Azure-Data-Explorer`. All the code and query samples listed in the book are available in our repository. Download the latest version of Git from `https://git-scm.com` if you have not already installed the command-line tools.

> **Important Note**
>
> When developing and cloning repositories, I create a development folder in my home directory. On Windows, this is `C:\Users\jason\development`. On macOS, this is `/Users/jason/development`. When referencing specific code examples, I will refer to the repository's parent directory as `${HOME}`, for example, `${HOME}/Scalable-Data-Analytics-with-Azure-Data-Explorer/Chapterxx/file.kql`.

Introducing the data analytics pipeline

Before diving into *ADX*, it is worth spending some time to understand the data analytics pipeline. Whenever I am learning something new that is large and complex in scope, such as data analytics, I break the topic down into smaller chunks to help with learning and measuring my progress. Therefore, an understanding of the various stages of the data analytics pipeline will help you understand how ADX takes raw data and generates reports and visuals as a result of our analytical tasks, such as time series analysis.

Figure 1.1 illustrates the stages of the data analytics pipeline required to take data from a data source, perform some analysis, and produce the result of the analysis in the form of a visual, such as tables, reports, and graphs:

Figure 1.1 – Data analytics pipeline

In the spirit of breaking a complex subject into smaller chunks, let's look at each stage in detail:

1. **Data**: The first step in the pipeline is the data sources. In *Chapter 4, Ingesting Data in Azure Data Explorer*, we will discuss the different types of data. For now, suffice it to say there are three different categories of data: **structured**, **semi-structured**, and **unstructured**. Data can range from structured, such as tables, to unstructured, such as free-form text.

2. **Ingestion**: Once the data sources have been identified, the data needs to be ingested by the pipeline. The primary purpose of the ingestion stage is to take the raw data, perform some **Extract-Transform-Load (ETL)** operations to format the data in a way that helps with your analysis, and send the data to the storage stage. The data can be ingested using tools and services such as Apache Kafka, Azure Event Hubs, and IoT Hub. *Chapter 4, Ingesting Data in Azure Data Explorer*, discusses the different ingestion methods, such as streaming versus batch, and demonstrates how to ingest data using multiple services, such as Azure Event Hubs and Azure Blob storage.

3. **Store**: Once ingested, ADX natively compresses and stores the data in a proprietary format. The data is then cached locally on the cluster based on the hot cache settings. The data is phased out of the cluster based on the retention settings. We will discuss these terms a little later in the chapter.

4. **Analyze**: At this stage, we can start to query, apply machine learning to detect anomalies, and predict trends. We will see examples of anomaly detection and trend prediction in *Chapter 7, Identifying Patterns, Anomalies, and Trends in Your Data*. In this book, we will perform most of our analysis in the ADX Web UI using **Kusto Query Language** (**KQL**).

5. **Visualize**: The final stage of the pipeline is *visualize*. Once you have ingested your data and performed your analysis, chances are you will want to share and present your findings. We will present our findings using the ADX Web UI's dashboards and Power BI.

In the next section, we will look at some of the services Azure provides for the different stages of the analytics pipeline.

Overview of Azure data analytics services

You may have noticed that I referenced a few of Azure's data services previously, and you may be wondering what they are used for. Although this book is about Azure Data Explorer, it is worth understanding what some of the common data services are, since some of the services, such as **Event Hubs** and **Blob storage**, will be discussed and used in later chapters.

To help map the different data services to the analytics pipeline, *Figure 1.2* illustrates an updated pipeline, with the Azure data services mapped to the respective pipeline stages:

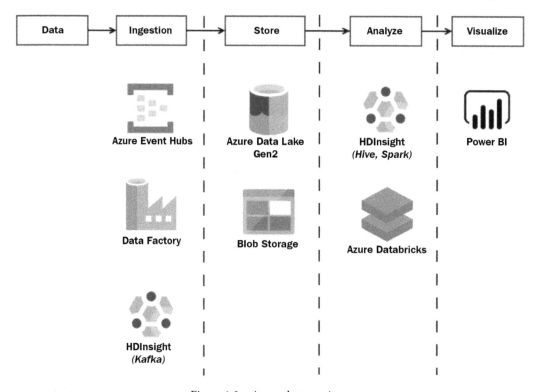

Figure 1.2 – Azure data services

Important Note

The list of services depicted in *Figure 1.2* is by no means an exhaustive list of Azure data analytics services. For a complete and accurate list, please see `https://azure.microsoft.com/en-us/services/#analytics`.

The following list of services is a short description of the services shown in *Figure 1.2*:

- **Event Hubs**: This is an event and streaming **Platform as a Service (PaaS)**. Event Hubs allows us to stream data, which we will demonstrate and use in *Chapter 4, Ingesting Data in Azure Data Explorer*.

- **Data Factory**: This is a PaaS service that allows us to transform data from one format to another. These transformations are commonly referred to as **Extract-Transform-Load (ETL)** and **Extract-Load-Transform (ELT)**.

- **HDInsight**: This is a PaaS service that appears twice in *Figure 1.2* and could technically appear in other stages. HDInsight is quite possibly one of the most misunderstood analytical services, with regard to what it does. HDInsight is a PaaS version of the **Hortonworks Hadoop** framework, which includes a wide range of ingestion, analytics, and storage services, such as **Apache Kafka**, **Hive**, **HBase**, **Spark**, and the **Hadoop Distributed File System (HDFS)**.

- **Azure Data Lake Gen2**: This is a storage solution based on Azure Blob storage that implements HDFS.

- **Blob Storage**: This is Azure's object storage service that all other storage services are based on.

- **Azure Databricks**: This is Azure's PaaS implementation of **Apache Spark**.

- **Power BI**: Technically not an Azure service, **Power BI** is a rich reporting product that is commonly integrated with Azure.

You may be wondering where ADX would fit in *Figure 1.2*. The answer is ingestion, store, analyze, and visualize. In the next section, you will learn how this is possible by understanding what Azure Data Explorer is.

What is Azure Data Explorer?

There is a good chance you have already used ADX to some degree without realizing it. If you have used **Azure Security Center**, **Azure Sentinel**, **Application Insights**, **Resource Graph Explorer**, or enabled diagnostics on your Azure resources, then you have used ADX. All these services rely on **Log Analytics**, which is built on top of ADX.

Like many tools and products, ADX was started by a small group of engineers circa 2015 who were trying to solve a problem. A small group of developers from Microsoft's Power BI team needed a high-performing big data solution to ingest and analyze their logging and telemetry data, and being engineers, they built their own when they could not find a service that met their needs. This resulted in the creation of Azure Data Explorer, also known as **Kusto**.

So, what is ADX? It is a fully managed, append-only columnar store big data service capable of elastic scaling and ingesting literally hundreds of billions of records daily!

Before moving onto the ADX features, it is important to understand what is meant by **PaaS** and the other cloud offerings referred to as *as a service*. Understanding the different cloud offerings will help with understanding what you and the cloud provider – in our case, Microsoft – are responsible for.

When you strip away the marketing terms, cloud computing is essentially a data center that is managed for you and has the same layers or elements as an on-premises data center, for example, hardware, storage, and networking.

Figure 1.3 shows the common layers and elements of a data center. The items in white are managed by you, the customer, and the items in gray are managed by the cloud provider:

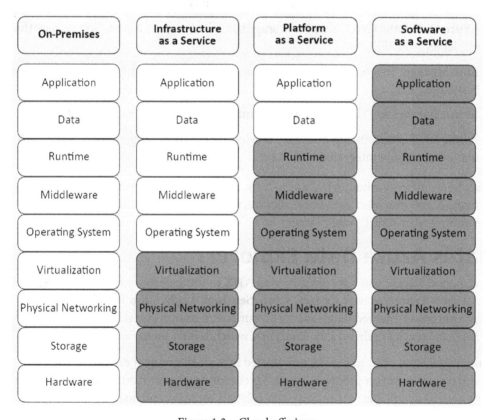

Figure 1.3 – Cloud offerings

In the case of on-premises, you are responsible for everything, from renting the building and ventilation to physical networking and running your applications. Public cloud providers offer three fundamental cloud offerings, known as **Infrastructure as a Service (IaaS)**, **Platform as a Service (PaaS)**, and **Software as a Service (SaaS)**. The provider typically offers a lot more services, such as Azure App Service, but these additional services are built on top of the aforementioned fundamental services.

In the case of ADX, which is a PaaS service, Microsoft manages all layers except the data and application. You are responsible for the data layer, that is, the data ingestion, and the application layer, that is, writing our KQL and creating dashboards.

ADX features

Let's look at some of the key features ADX provides. Most of the features will be discussed in detail in later chapters:

- **Low-latency ingestion and elastic scaling**: ADX nodes are capable of ingesting structured, semi-structured, and unstructured data up to speeds of 200 MBps (megabytes per second). The vertical and horizontal scaling capabilities of ADX enable it to ingest petabytes of data.

- **Time series analysis**: As we will see in *Chapter 7, Identifying Patterns, Anomalies, and Trends in Your Data*, ADX supports near real-time monitoring, and combined with the powerful KQL, we can search for anomalies and trends within our data.

- **Fully managed (PaaS)**: All the infrastructure, operating system patching, and software updates are taken care of by Microsoft. You can focus on developing your product rather than running a big data platform. You can be up and running in three steps:

  - Create a cluster and database (more details in *Chapter 2, Building Your Azure Data Explorer Environment*).

  - Ingest data (more details in *Chapter 4, Ingesting Data in Azure Data Explorer*).

  - Explore your data using KQL (more details in *Chapter 5, Introducing the Kusto Query Language*).

- **Cost-efficient**: Like other Azure services, Microsoft provides a pay-as-you-consume model. For more advanced use cases, there is also the option of purchasing reserved instances, which require upfront payments.

- **High availability**: Microsoft provides an uptime SLA of 99.9% and supports Availability Zones, which ensures your infrastructure is deployed across multiple physical data centers within an Azure region.

- **Rapid ad hoc query performance**: Due to some of the architecture decisions that are discussed in the next section, ADX is capable of querying billions of records containing structured, semi-structured, and unstructured data, returning results within seconds. ADX is also designed to execute distributed queries across multiple clusters, which we will see later in the book.

- **Security**: We will be covering security in depth in *Chapter 10*, *Azure Data Explorer Security*. For now, suffice it to say that ADX supports both encryption at rest and in transit, **role-based access control** (**RBAC**), and allows you to restrict public access to your clusters by deploying them into **virtual private networks** (**VPNs**) and block traffic using **network security groups** (**NSGs**).

- **Enables custom solutions**: Allows developers to build analytics services on top of ADX.

If you are familiar with database products such as MySQL, MS SQL Server, and Azure SQL, then the core components will be familiar to you. ADX uses the concept of clusters, which can be considered equivalent to Azure SQL Server and are essentially the compute or virtual machines. Next, we have databases and tables; these concepts are the same as a SQL database.

Figure 1.4 shows the hierarchical structure that is shown in the Data Explorer UI. In this example, **help** is the ADX cluster and **Samples** is in the database, which contains multiple tables such as **US_States**:

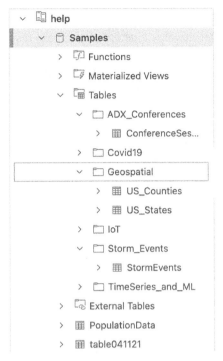

Figure 1.4 – Cluster, database, and tables hierarchy

A cluster or SQL server can host multiple databases, which in turn can contain multiple tables (see *Figure 1.4*). We will discuss tables in *Chapter 4, Ingesting Data in Azure Data Explorer*, when we will demonstrate how to create tables and data mappings.

Introducing Azure Data Explorer architecture

PaaS services are great because they allow developers to get started quickly and focus on their product rather than managing complex infrastructure. Being fully managed can also be a disadvantage, especially when you experience issues and need to troubleshoot, and as engineers, we tend to be curious and want to understand how things work.

As depicted in *Figure 1.5*, ADX contains two key services, the data management service and the engine service. Both services are clusters of compute resources that can be automatically or manually scaled horizontally and vertically. At the time of writing, Microsoft recently announced their V3 engine in March 2021, which contains some significant performance improvements:

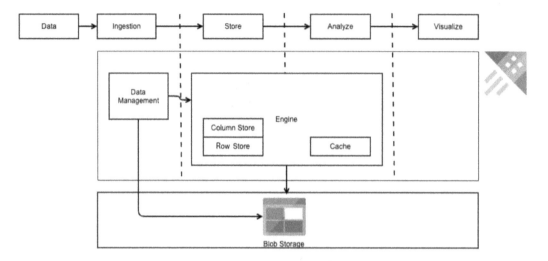

Figure 1.5 – Azure Data Explorer architecture

Now, let's learn more about the data management and the engine service depicted in the preceding diagram:

- **Data management service**: The **data management service** is responsible primarily for metadata management and managing the data ingestion pipelines. The data management service ensures data is properly ingested and sent to the engine service. Data that is streamed to our cluster is sent to the row store, whereas data that is batched is sent to the column stores.

- **Engine service**: The **engine service**, which is a cluster of compute resources, is responsible for processing the ingested data, managing the hot cache and the long-term storage, and query execution. Each engine uses its local SSD as the hot cache and ensures the cache is used as much as possible.

ADX is often referred to as an append-only analytics service, since the data that is ingested is stored in immutable shards and each shard is compressed for performance reasons. **Data sharding** is a method of splitting data into smaller chunks. Since the data is immutable, the engine nodes can safely read the data shards, knowing they do not have to worry about other nodes in the cluster making changes to the data.

Since the storage and the compute are decoupled, ADX can scale the cluster both vertically and horizontally without worrying too much about data management.

This brief overview of the architecture only scratches the surface; there are a lot more tasks happening, such as indexing columns and maintenance of the indexes. Having an overview helps appreciate what ADX is doing under the hood.

> **Important Note**
>
> I recommend reading the Azure Data Explorer white paper `https://azure.microsoft.com/mediahandler/files/resourcefiles/azure-data-explorer/Azure_Data_Explorer_white_paper.pdf` if you are interested in learning more about the architecture.

Azure Data Explorer use cases

Whenever someone asks what they should focus on when learning how to use Azure, I immediately say KQL. I use KQL daily, from managing cost and inventory to security and troubleshooting. It is not uncommon for relatively small environments to generate hundreds of GB of data per day, such as infrastructure diagnostics, **Azure Resource Manager** (**ARM**) audit logs, user audit logs, application logs, and application performance data. This may seem small in the grand scheme of things when, in 2021, we are generating quintillion bytes of data per day. But it is still enough data to require dedicated services such as ADX to analyze the data.

IoT monitoring and telemetry

Look around at your environment: how many appliances and devices can you see that are connected to the network? I see light bulbs, sensors, thermostats, and fire alarms, and there are billions of **Internet of Things** (**IoT**) devices in the world, all of which are constantly generating data. Together with Azure's IoT services, ADX can ingest the high volumes of data and enable us to monitor our *things* and perform complex time series analysis, so that we can identify anomalies and trends in our data.

Log analysis

Imagine this scenario: you have just performed a lift-and-shift migration to Azure for your on-premises product, and since the application is not a true cloud-native solution, you are constrained by which Azure services you can use, such as load balancing. **Azure Application Gateway**, which is a load-balancing service, supports cookie-based session affinity, and the cookies are completely managed by Application Gateway. The application we migrated to Azure required specific values to be written in the cookie, and this is not possible with the current version of Application Gateway, so we used HAProxy running on Linux virtual machines. The security team requires all products to only support TLS 1.2 and above. The problem is that not all of our clients support TLS 1.2, and if we simply disabled TLS 1.0 and 1.1, we would essentially break the service for those clients, which we do not want to do. Add to the equation the server-side product, which is distributed across 15 Azure Regions worldwide with each region containing hundreds of the HAProxy servers with no central logging! How can we analyze all this data to identify the clients that are not using TLS 1.2? The answer is Kusto.

We ingested the HAProxy log files and used KQL to analyze the log files and capture insights on TLS versioning and cipher information in seconds. With the queries, we were able to build near real-time dashboards for the support teams so they could reach out to clients and inform them when they would need to upgrade their software. With these insights, we were able to coordinate the TLS deprecation activities and execute them with no customer impact.

Most of the examples in this book focus on logging scenarios, and in *Chapter 7, Identifying Patterns, Anomalies, and Trends in Your Data*, we will learn about ADX's time series analysis features to identify patterns, anomalies, and trends in our data.

Running your first query

In this section, we are going to clone the Git repository, connect to an example ADX cluster called `https://help.kusto.windows.net`, which is provided by Microsoft, execute our first KQL query, and generate a bar chart showing population data per state in the US:

1. If you are using Windows, open a new **PowerShell** terminal; if you are using macOS, then open a new **Terminal** or a **Shell** if you are using Linux. Clone the accompanying Git repository by typing `git clone https://github.com/PacktPublishing/Scalable-Data-Analytics-with-Azure-Data-Explorer.git`.

2. Open your browser and go to `https://dataexplorer.azure.com`. The URL will take you to the ADX UI, the interactive environment for querying your data. Do not worry about the panes and layout. The ADX UI will be discussed in more detail in *Chapter 3, Exploring the Azure Data Explorer UI*. If you are prompted, log in with your Microsoft account. If your browser gets stuck in a refresh/redirect loop, please ensure you have allowed third-party cookies and check your *SameSite* cookie settings. You need to allow cookies from `portal.azure.com` and `dataexplorer.azure.com`. This issue is rare and only happened once when using Safari on macOS. Microsoft added a nice convenient feature whereby the web UI automatically connects to the help cluster (`https://help.kusto.windows.net/`). You can skip *steps 3 and 4* if you see the help cluster in the cluster pane.

3. Once you are logged in, the next step is to connect to an ADX cluster. Microsoft provides an example cluster, which we will be using throughout the book alongside the cluster we will create in *Chapter 2, Building Your Azure Data Explorer Environment*.

    Click **Add Cluster**, as shown in *Figure 1.6*, and enter `https://help.kusto.windows.net` as the connection URL. The **Display Name** field will be populated with the cluster name, in this instance, **help**.

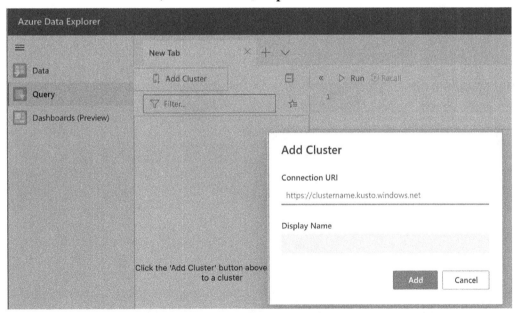

Figure 1.6 – Connecting to ADX clusters

4. Once connected, the **help** cluster will appear in the cluster pane, which is located below the **Add Cluster** button, as shown in *Figure 1.7*:

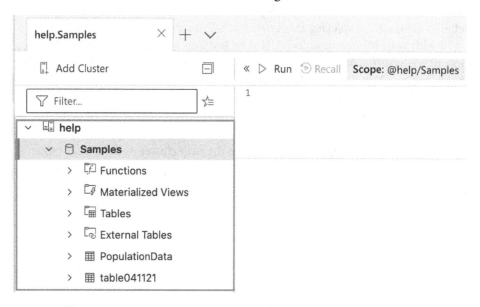

Figure 1.7 – Connected to the help cluster

5. Expand the cluster and click the **Samples** database to set the scope to **@help/Samples**, as shown in *Figure 1.8*:

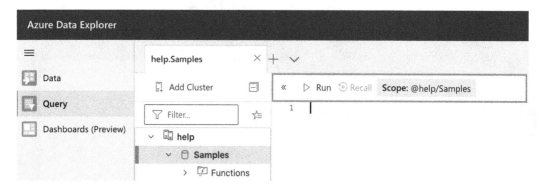

Figure 1.8 – Expand the cluster and set the scope

6. Next, click **File | Open**, as shown in *Figure 1.9*, and open `${HOME}/Scalable-Data-Analytics-with-Azure-Data-Explorer/Chapter01/first-query.kql`:

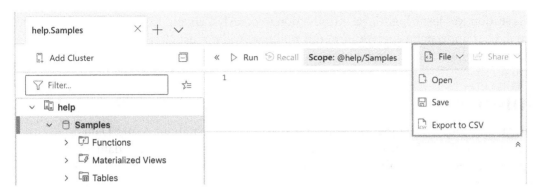

Figure 1.9 – Open query in ADX

7. Finally, click **Run**, and you should see the population data, as shown in *Figure 1.10*. Even though the dataset is relatively small, ADX sorted the states by population in descending order and rendered the result as a bar chart in under 1 second:

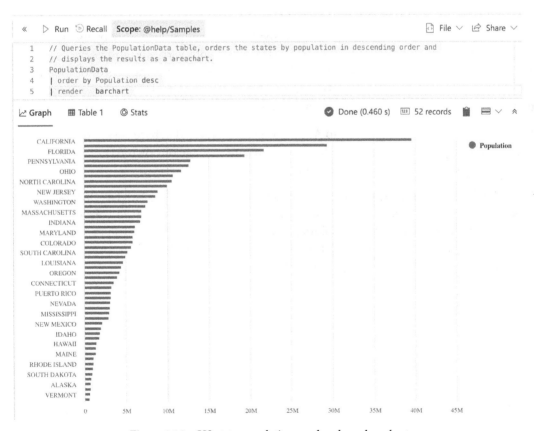

Figure 1.10 – US state population rendered as a bar chart

This section was just a high-level introduction to the ADX UI and KQL. Don't worry if you did not understand everything at this point; we will discuss all the topics in more detail throughout the remainder of the book.

Summary

Congratulations on completing your first steps in learning about Azure Data Explorer! In this chapter, you learned about the different stages of the data analytics pipeline. Understanding the stages of the pipeline helps simplify your ability to comprehend the workflow of taking raw data and performing analysis on the data and visualizing your findings.

We then introduced some of the popular Azure data analytics services and mapped them to the different stages of the data analytics pipeline. Some of the services, such as Event Hubs, will be used in later chapters to ingest data into our own ADX databases.

We then learned what ADX is, what the main features are, and briefly looked at the ADX architecture to understand how ADX provides excellent performance by using both column stores and row stores, and how ADX scales both vertically and horizontally efficiently by implementing one of the fundamental Azure design principles of decoupling compute and storage. We then discussed some of the use cases of ADX that we will use throughout this book, such as time series analysis.

Finally, we learned how to connect to ADX clusters and query databases using the ADX UI. In the next chapter, we will learn how to create and manage our own ADX clusters and databases using the Azure portal, PowerShell, and the Azure CLI.

Before moving on to the next chapter, try modifying `${HOME}/Scalable-Data-Analytics-with-Azure-Data-Explorer/Chapter01/first-query.kql` and display an area chart. The solution can be found at `${HOME}/Scalable-Data-Analytics-with-Azure-Data-Explorer/Chapter01/population-areachart.kql`. What other types of charts can you render?

Additionally, here is some information you should know. The Azure Data Explorer UI supports a feature known as **IntelliSense**, as shown in *Figure 1.11*. IntelliSense provides code completion and hints when you are writing your queries, so you do not need to worry about memorizing all the keywords:

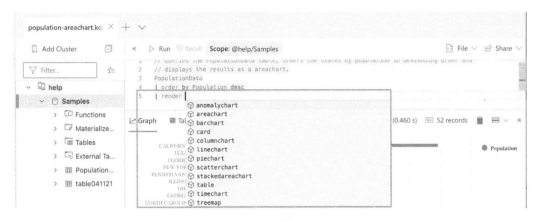

Figure 1.11 – IntelliSense features

We will be using IntelliSense throughout this book when using both **Visual Studio Code** and the **Azure Data Explorer Web UI**. Visual Studio Code will be used for editing our scripts and ARM templates, and the Azure Data Explorer Web UI is where we will execute most of our KQL queries.

2
Building Your Azure Data Explorer Environment

In the previous chapter, we introduced the data analytics pipeline, **Azure Data Explorer** (**ADX**), and executed our first **Kusto Query Language** (**KQL**) query on a publicly available demo cluster provided by **Microsoft**.

In this chapter, we will assume you have just created a new Azure account and begin by creating a new subscription. Once we have a subscription, we can start creating Azure resources, such as **Azure Cloud Shell** and ADX instances.

Next, we will introduce you to Cloud Shell, provision our first Cloud Shell instance, and discover one of the least known but extremely useful features, the lightweight code editor.

Then, we will create our first **ADX clusters and database** via the Azure portal, introduce the concept of **Infrastructure as Code** (**IaC**) and discuss some of the benefits and why IaC should be your preferred method for managing infrastructure on Azure.

Next, we will use Cloud Shell to recreate our ADX clusters and databases using PowerShell, and finally, we will introduce **Azure Resource Manager (ARM) templates**, demonstrate how to reuse them, and how to deploy an ADX cluster and database using our ARM templates.

In this chapter, we are going to cover the following main topics:

- Creating an Azure subscription

- Introducing Azure Cloud Shell

- Creating and configuring ADX instances in the Azure portal

- Introducing Infrastructure as Code

- Creating and configuring ADX instances with PowerShell

- Creating and configuring ADX instances with ARM templates

Technical requirements

If you do not already have an Azure account, head over to `https://azure.microsoft.com/en-us/free/search/` and sign up. Microsoft provides 12 months of popular free services and $200 credit at the time of writing.

The code examples for this chapter can be found in the `Chapter02` folder of our repo: `https://github.com/PacktPublishing/Scalable-Data-Analytics-with-Azure-Data-Explorer.git`.

For the PowerShell examples, we are going to use Azure Cloud Shell rather than installing and configuring the tools locally on our machines.

Table 2.1 shows the **PowerShell cmdlet** versions used for the PowerShell examples:

Component	Version
Az	6.2.1
Az.Kusto	2.0.0

Table 2.1 – PowerShell cmdlet version info

All the code examples, both ARM templates and PowerShell, have been tested on Azure Cloud Shell, macOS, and Windows 10.

Creating an Azure subscription

Once you have created your Azure account and logged in, you will see the **Welcome to Azure** page. You will be informed that you do not have a subscription.

There are different types of subscriptions, each with different constraints and conditions, such as free credit and **Service Level Agreements** (**SLAs**) for Azure services. For our purpose, we will use the **Free Trial**. If your trial expires, then you can upgrade the subscription to **Pay-As-You-Go**. For a complete list of Azure subscriptions, please see https://azure.microsoft.com/en-us/support/legal/offer-details/.

Before you can create any resources in Azure, you need to create a subscription. Subscriptions in Azure are logical containers, primarily used for billing. Any costs you incur will be assigned to your subscription. It is possible to have multiple subscriptions, but we only need one for our purposes.

As shown in *Figure 2.1*, the default behavior is to collapse the **Azure portal** menu. To view the menu, click the hamburger icon in the top-left corner.

Figure 2.1 – Azure portal

Let's create our first Azure subscription so we can begin to deploy our Cloud Shell and ADX clusters:

1. From the portal menu, click **All services** and search for **Subscriptions**, as shown in *Figure 2.2*:

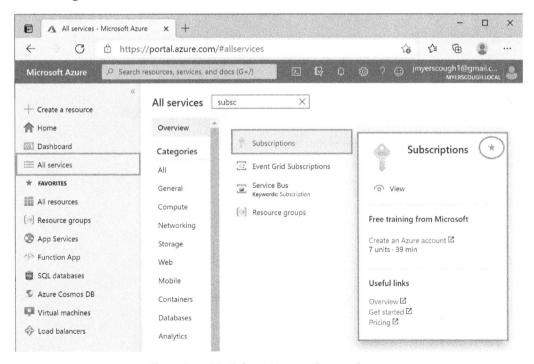

Figure 2.2 – Pin Subscriptions to the portal menu

> **Note**
> You can pin services to the portal menu by hovering over a service and clicking the star that appears on the service dialog, as shown in *Figure 2.2*.

2. Once you click **Subscriptions**, the **Subscriptions** blade appears with empty fields, as shown in *Figure 2.3*. When you click **+ Add**, you will be taken to the subscription blade.

Figure 2.3 – Subscriptions blade

3. As shown in *Figure 2.4*, select **Free Trial**. You will then be redirected to a signup page, where you will be required to provide a credit card and a mobile phone number for verification purposes.

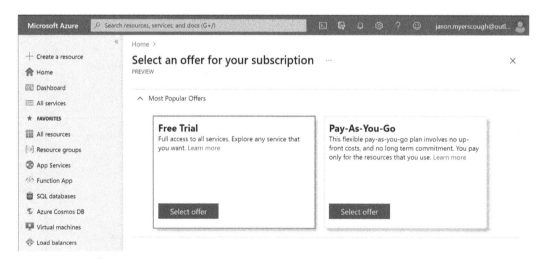

Figure 2.4 – Activate your free trial

> **Note**
>
> Your credit card will not be charged, but once your trial expires, you will need to upgrade your subscription to the pay-as-you-go model. Once you upgrade your subscription, your credit card will be charged. In *Chapter 12, Cost Management in Azure Data Explorer*, we will discuss cost management and demonstrate how to set budgets and alerts on your spending.

4. Once you complete this process, a new subscription called **Free Trial** will appear on the **Subscriptions** blade (see *Figure 2.5*).

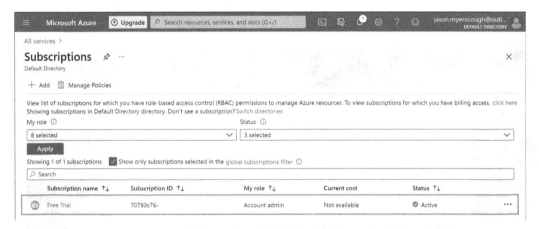

Figure 2.5 – Trial subscription

With our newly created subscription, we can start creating resources in Azure such as Cloud Shell and ADX instances. In the next section, *Introducing Azure Cloud Shell*, we will provision Azure Cloud Shell so we can start using PowerShell to deploy ADX clusters using the Kusto cmdlets and ARM templates.

Introducing Azure Cloud Shell

One of the most convenient services, which I use daily, is Azure Cloud Shell. Azure Cloud Shell is an interactive command shell that supports **Bash** and **PowerShell**. Rather than installing the PowerShell Az module locally, we will use Azure Cloud Shell for the examples in this book. Azure Cloud Shell also includes a lightweight code editor that we will see later when we examine the code examples for this chapter.

Azure Cloud Shell requires a storage account, which can be created the first time you try to open Cloud Shell. Azure Cloud Shell can be accessed via `https://shell.azure.com` or by clicking the Azure Cloud Shell icon on the Azure portal's header menu (see *Figure 2.6*).

Figure 2.6 – Cloud Shell icon

The first time you open Azure Cloud Shell, you will be prompted to select either **Bash** or **PowerShell** as your default shell (see *Figure 2.7*). Select **PowerShell**. You can switch shells at any time.

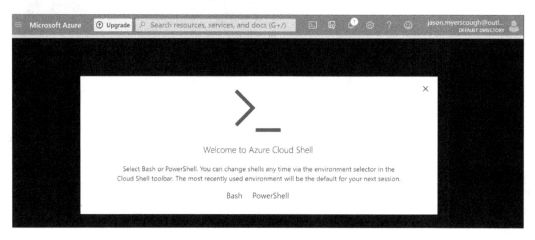

Figure 2.7 – Initializing Cloud Shell

Next, you will be prompted to specify a storage account. If you already have a storage account, you can click **Show advanced settings** and select the account. The advanced settings also allow you to create a new storage account with a specific name. In our case, you can simply click **Create storage** and Azure will take care of provisioning a storage account for you.

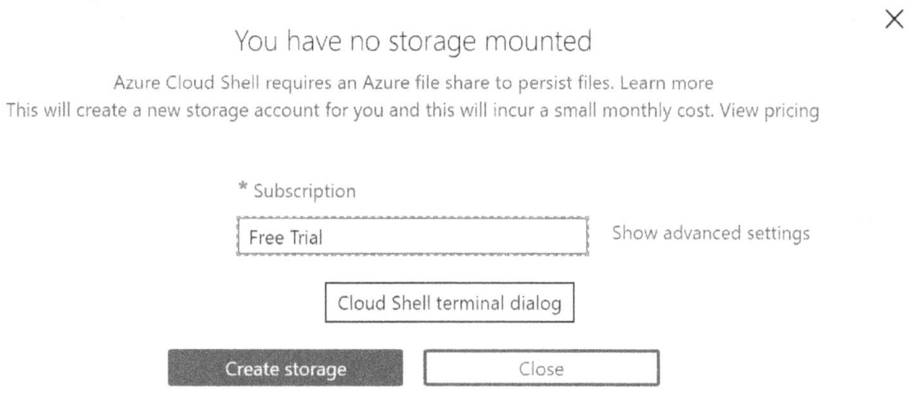

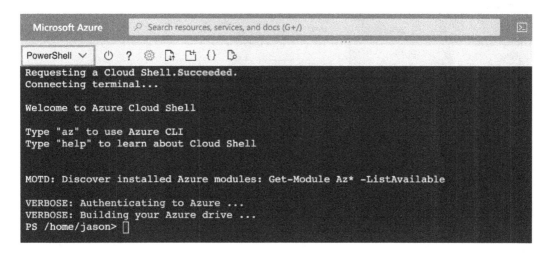

Figure 2.8 – Create a storage account for Cloud Shell

Once the storage account is provisioned, you can start using your shell. As shown in *Figure 2.9*, the drop-down menu in the top-left corner can be used to toggle between **PowerShell** and **Bash**.

Figure 2.9 – PowerShell in the Azure portal

Now that we have our subscription and tools configured, we can now proceed with creating our ADX cluster and database. The next section will demonstrate how to create ADX clusters and databases using the Azure portal, PowerShell, and ARM templates.

Creating and configuring ADX instances in the Azure portal

As mentioned in the previous chapter, one of the benefits of **Platform as a Service (PaaS)** solutions is that they abstract a lot of the operational work away from you, allowing you to focus more on using the product rather than operating it. At a high level, there are four steps to get up and running:

1. Creating a cluster.
2. Creating a database.
3. Ingesting data.
4. Querying your data.

To create a cluster, you need to complete the following steps:

1. In the portal menu, click **All services** and search for data explorer, as shown in *Figure 2.10*.

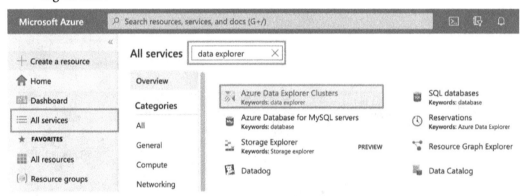

Figure 2.10 – Azure Data Explorer Clusters

2. Click **Azure Data Explorer Clusters** to display the ADX blade. Create a new cluster by clicking **+ Create**, as shown in *Figure 2.11*.

Figure 2.11 – Create a new cluster

The **Create an Azure Data Explorer Cluster** blade displays a wizard with different configuration stages.

As shown in *Figure 2.12*, the configuration wizard is broken up into the following categories: **Basics**, **Scale**, **Configurations**, **Security**, **Network**, **Diagnostic settings**, **Tags**, and finally, **Review + create**.

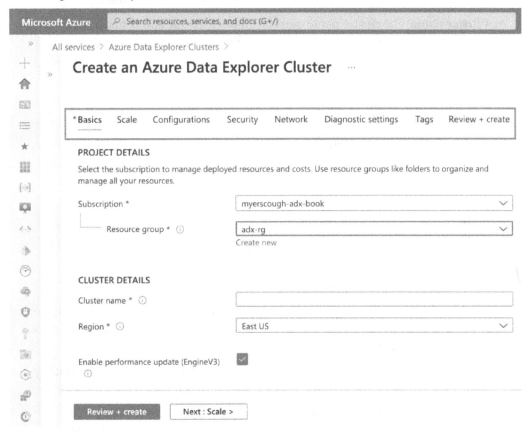

Figure 2.12 – Create an Azure Data Explorer Cluster blade

Let's walk through the wizard and look at each category in detail:

- The **Basics** section of the configuration is where we can specify the subscription, resource group, name, location, engine version, high availability, and virtual machine size of our cluster:

  - **Subscription**: As mentioned earlier, the subscription is a logical container for resources, which is used to track monetary costs. Select your **Free Trial** subscription.

- **Resource group**: Resource groups are like subscriptions in terms of being a logical container for resources. All Azure resources must reside in a resource group, and you typically assign resources to a resource group that has the same life cycle. This allows you to easily create, update, and destroy resources. Click **Create new** and enter a name for your resource group, for example, `adx-rg`. As a naming convention, I like to append `-rg` to the end of resource group names.

- **Cluster name**: The cluster name must be globally unique and contain between 4 and 22 alphanumeric characters. Enter a cluster name, for example, `adxmyerscough`.

- **Region**: The region refers to the Azure Region where your resources will be deployed. I recommend selecting a region that is close to you, for example, I selected **West Europe**.

- **Enable performance update (EngineV3)**: As mentioned in the previous chapter, at the time of writing, EngineV3 has been recently released and contains some significant performance improvements. Enable this option.

> **Note**
> In *Chapter 12*, *Cost Management in Azure Data Explorer*, we will discuss how to select the right workload for your use case. In the meantime, we will use the **Dev/test** type.

- **Workload**: The workload specifies which workload type the cluster should be optimized for. The standard SKUs can be optimized for either storage or compute. For this example, select **Dev/test**.

- **Size**: The size refers to the number of CPU cores. The **Dev/test** SKU does not allow us to configure the size, whereas the standard SKUs do.

- **Compute specifications**: The compute specifications refer to the virtual machine type. At the time of writing, the **Dev/test** SKU is limited to two SKUs and provides **Dev(NoSLA)_Standard_E2a_v4** and **Dev(NoSLA)_Standard_D11_v2**. We will discuss the different VM SKUs in *Chapter 12*, *Cost Management in Azure Data Explorer*.

- **Availability zones**: The Availability Zones improve high availability by ensuring your resources are deployed in multiple physical data centers within a metropolitan area. For now, leave this option as **(none)**, as shown in *Figure 2.13*:

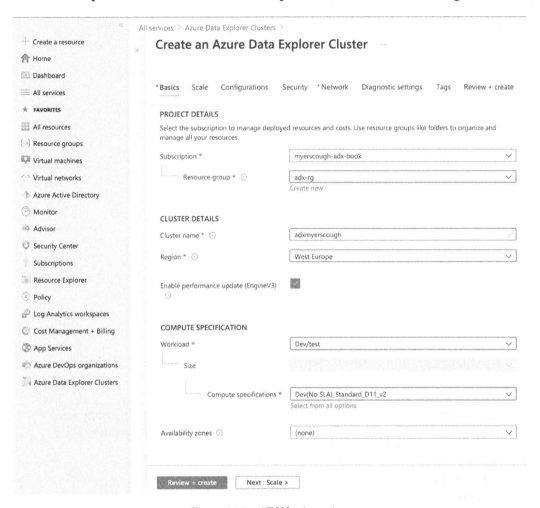

Figure 2.13 – ADX basic settings

- The **Scale** section of the configuration is where we can specify our horizontal scaling requirements. We will discuss scaling in detail in *Chapter 12, Cost Management in Azure Data Explorer*:

- For now, keep the scaling method to **Manual scale** and **Instance count** to **1**, as shown in *Figure 2.14*:

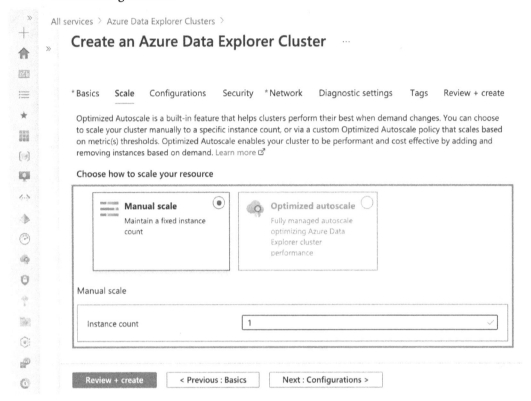

Figure 2.14 – Scaling options

- The **Configurations** section is where we configure our streaming ingesting and purging requirements. We will cover data ingestion in *Chapter 4, Ingesting Data in Azure Data Explorer*:

  - **Streaming ingestion**: This option provides low latency between the point your data is ingested and the point where you can query the data. Keep this option disabled for now; it can be changed at a later point.

  - **Enable purge**: As mentioned in *Chapter 1, Introducing Azure Data Explorer*, ADX is an append-only analytics service. To meet compliance requirements, ADX does support the ability to purge data, but this feature needs to be explicitly enabled. We can go ahead and keep this feature disabled since we do not need to be GDPR compliant for our purposes. We will discuss this option in *Chapter 11, Performance Tuning in Azure Data Explorer*, and cover some of the performance impacts that could be introduced if you enable this feature. For now, select **Off**, as shown in *Figure 2.15*.

- **Auto-Stop cluster**: The auto stop cluster setting stops our cluster after a period of inactivity. Once you have created your cluster, you will receive recommendations from Azure Advisor to stop your cluster. If you do not act on these recommendations, your inactive cluster will be stopped after 5 days if you have negligible amount of data on the cluster. If you have data on your cluster, the cluster will be stopped after 10 days. In the spirit of managing costs, keep **Auto-Stop cluster** to **On**. Azure Advisor will be discussed in more detail in *Chapter 12, Cost Management in Azure Data Explorer*. The following screenshot shows these options:

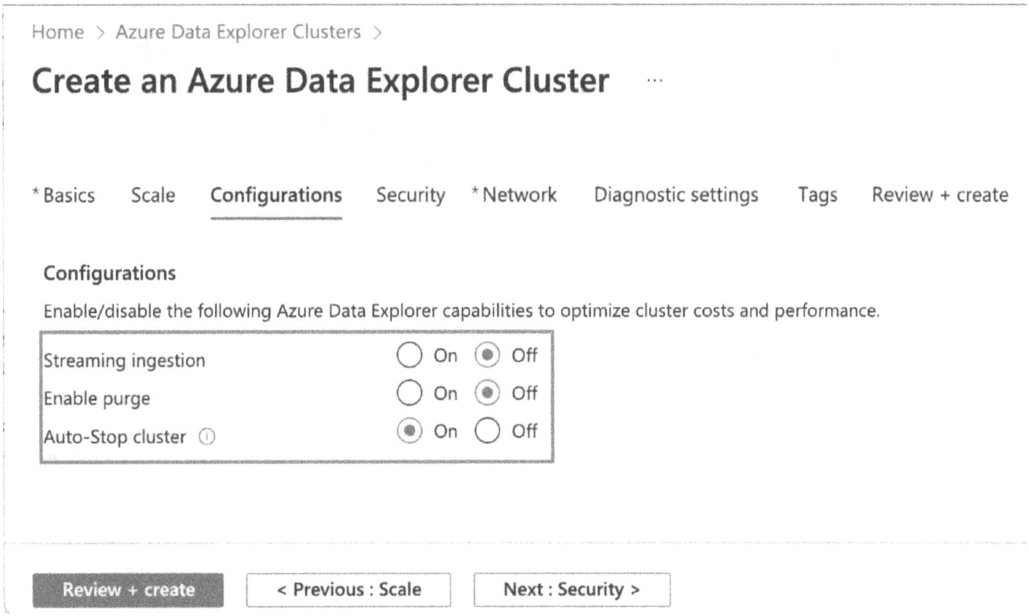

Home > Azure Data Explorer Clusters >

Create an Azure Data Explorer Cluster   ···

* Basics   Scale   **Configurations**   Security   * Network   Diagnostic settings   Tags   Review + create

Configurations

Enable/disable the following Azure Data Explorer capabilities to optimize cluster costs and performance.

Streaming ingestion	○ On ◉ Off
Enable purge	○ On ◉ Off
Auto-Stop cluster ⓘ	◉ On ○ Off

Review + create   < Previous : Scale   Next : Security >

Figure 2.15 – Streaming and purging settings

- The **Security** section is where we can configure additional encryption settings and identity management. We will cover data ingestion in *Chapter 4, Ingesting Data in Azure Data Explorer*:

  - **Enable Double Encryption**: By default, ADX encrypts your data at the service level using 256-bit AES encryption. The **Enable Double Encryption** option provides the ability to enable an extra layer of encryption at the infrastructure layer. Each layer is encrypted with a different key, so an attacker would need to gain access to both keys to decrypt your data. Service-level encryption supports customer-managed keys, whereas infrastructure-level encryption only supports Microsoft-managed keys. For our purposes, we can go ahead and keep this option turned **Off**.

> **Note**
>
> **Enable Double Encryption** cannot be changed after the cluster has been created. Therefore, if you require double encryption, you must enable this feature when creating your instance.

- **Define tenants permissions**: By default, ADX only permits users from your Azure Active Directory tenant. Since a detailed discussion regarding Azure Active Directory tenants is beyond the scope of this book, please keep the default setting: **My tenant only**.

- **Identity**: There are two identity options, **System Assigned Identity** and **User Assigned Identity**, commonly known as managed identities. Both identity types are essentially **Azure Active Directory** service principals. Rather than dealing with credentials in your code and applications to authenticate and access Azure resources, managed identities allow you to manage access to your resources using **Role-Based Access Control** (**RBAC**). We will discuss managed identity in *Chapter 10*, *Azure Data Explorer Security*. For now, keep both options set to **Off**.

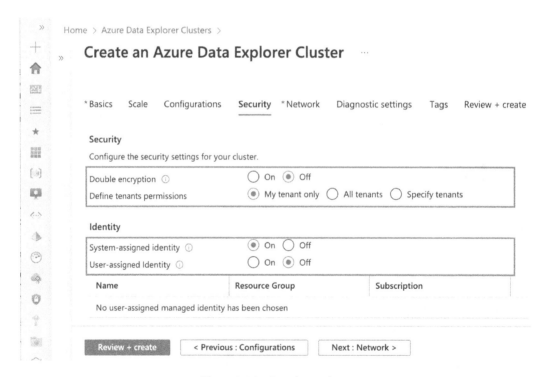

Figure 2.16 – Security settings

- The **Network** section is where we configure where we can deploy our ADX cluster in a **virtual network**. We will cover this in *Chapter 10, Azure Data Explorer Security*. By deploying our ADX cluster in a virtual network, we can use **Network Security Groups (NSGs)** to restrict access to our cluster:

  - For now, keep **Deploy in a virtual network?** set to **Off** (see *Figure 2.17*) and click **Next : Diagnostic settings >**.

All services > Azure Data Explorer Clusters >

Create an Azure Data Explorer Cluster …

| *Basics | Scale | Configurations | Security | *Network | Diagnostic settings | Tags | Review + create |

Network (West Europe)

Configure the network settings for your cluster

Deploy in a virtual network? ⓘ   ○ On   ◉ Off

Subscription   myerscough-adx-book

Query Public IP ⓘ

Create new

Data ingestion Public IP ⓘ

Review + create   < Previous : Security   Next : Diagnostic settings >

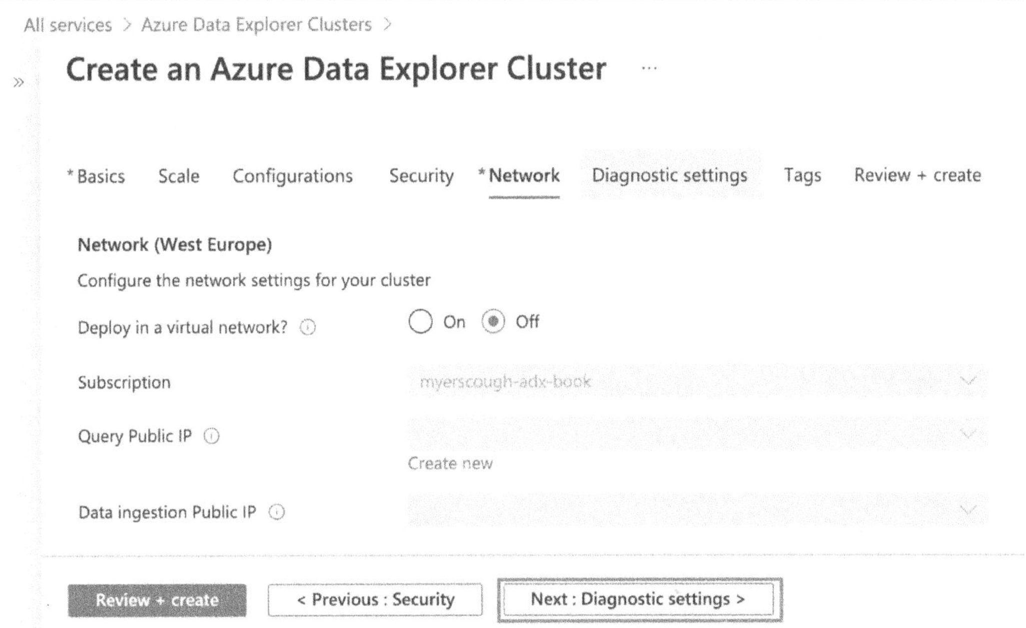

Figure 2.17 – Network settings

- The **Diagnostic settings** section allows us to enable and send ADX cluster's telemetry to a Log Analytics workspace. This telemetry is very useful when troubleshooting and performance tuning. For now, we can keep diagnostic settings turned off and in *Chapter 9, Monitoring and Troubleshooting Azure Data Explorer*, we will enable diagnostics and demonstrate how to troubleshoot a data ingestion problem. Click **Next: Tags >** to proceed to the next section.

- The **Tags** section allows us to assign metadata to our resources. One common use for tags is to assist with cost management, for example, we can create a tag called `cost center` to identify which department owns the resources. We can then use the tags to filter and drill down into our billing information.

  For our purposes, we can skip adding any tags. Click **Next: Review + create >** to proceed to the final step.

- The **Review + create** section provides an overview of our configuration that we can review before initiating the deployment.

  Once you have reviewed the deployment, click **Create** to start the deployment. The cluster can take on average 10-15 minutes to be deployed and get ready for use.

Once the cluster has been created, the next step is to create a database. As shown in *Figure 2.18*, click **Go to resource**:

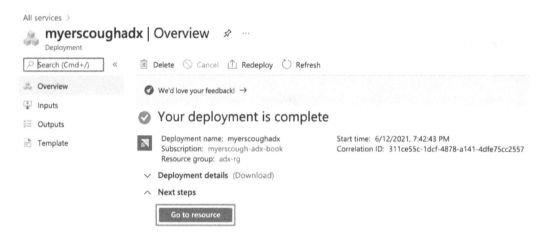

Figure 2.18 – ADX cluster deployment

The remaining step will walk through creating a database.

Click **Add database** to create your first ADX database. There are only three properties to set when creating an ADX database:

- **Database name**: Unlike the cluster name, the database name does not need to be globally unique, but it does need to be unique within the scope of your cluster. Remember, it is possible to have multiple databases on a cluster. For the name, enter adxdemo.

- **Retention period (in days)**: The retention period refers to how long the data is available to be queried. Once the retention period is exceeded, the data will be deleted and no longer available. For now, keep the retention as the default **365** days (1 year).

- **Cache period (in days)**: If you recall from *Chapter 1, Introducing Azure Data Explorer*, ADX uses two types of storage. **Hot cache**, which is the local SSD drives on the engine nodes, and cold storage, which is **blob storage**. The cache period refers to how long data is kept in the cache (local SSD storage). Once the cache period is exceeded, the data will be flushed from the cache. Keep the cache period as the default **31** days and click **Create**.

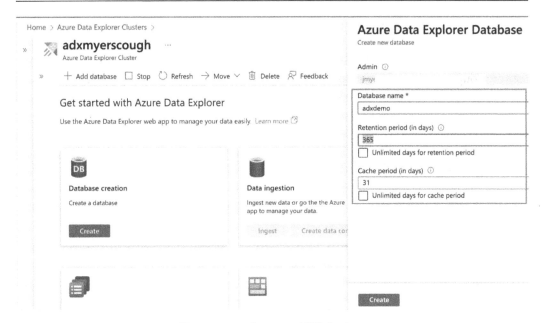

Figure 2.19 – Create an ADX database

Congratulations! You have created your first *ADX cluster and database*. The third step is to configure data ingestion, which we will perform in *Chapter 4, Ingesting Data in Azure Data Explorer*. If you are following the rest of this chapter, you can go ahead and delete the resource group we created since we are going to recreate it and the ADX cluster using PowerShell and ARM templates. Deleting the resource group will result in the deletion of the ADX cluster. In *Creating and configuring ADX instances with PowerShell* section, we will recreate the cluster using PowerShell and ARM templates.

As shown in *Figure 2.20*, click **Resource groups** and then click **adx-rg** to open the **Resource groups** blade:

Figure 2.20 – Delete the adx-rg resource group

To delete the resource group, click **Delete resource group**, as shown in *Figure 2.21*. Before you can delete the resource group, you will be prompted to enter the name of the resource group you wish to delete to reduce the risk of accidental deletion. Once you enter the resource group name, the **Delete** button will be enabled. Go ahead and click **Delete** to remove your resource group and ADX cluster.

Figure 2.21 – Delete adx-rg and the ADX cluster

The deletion time can vary, depending on how busy the Azure service is and what Azure region you are using. While writing, I experienced intervals ranging from 5 minutes up to 25 minutes for deleting my resources. Once you initiate the deletion, you can track the progress in the notifications sidebar, as shown in *Figure 2.22*:

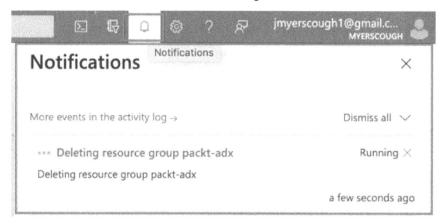

Figure 2.22 – Operation notifications

Before redeploying our ADX clusters using PowerShell and ARM templates, we will spend some time in the next section, *Introducing Infrastructure as Code (IaC)*, introducing IaC and discussing the benefits.

Introducing Infrastructure as Code

The Azure portal is a good place to start learning about different Azure resources. The simple user interface and rich documentation enables us to get up and running quickly. One of the disadvantages of using the portal is the deployment process is not consistently repeatable. Every time you want to deploy a resource, you must step through the wizard, which increases the chances of human error, such as specifying the incorrect location or even the incorrect subscription if you manage multiple subscriptions.

The preferred method for deploying and managing infrastructure is called **Infrastructure as Code** (**IaC**). IaC allows us to declare our infrastructure as code, giving us all the benefits of software development, for example, CI/CD, source control, code reviews, versioning, and so on. Since our infrastructure is in code, we can safely and reliably deploy our infrastructure consistently by deploying our code. This ability to deploy your infrastructure consistently is known as idempotency. **Idempotency** is a common IaC and **configuration management** term that means you will always get the same result or output, no matter how many times you deploy your code.

Azure provides a service known as the **Azure Resource Manager** (**ARM**), which is the deployment and management service that allows us to create, update, and delete our resources. Microsoft provides multiple ways to interact with ARM, for example, the Azure portal, Azure PowerShell, SDKs (Python, C#, and so forth), and the direct REST API (see *Figure 2.23*).

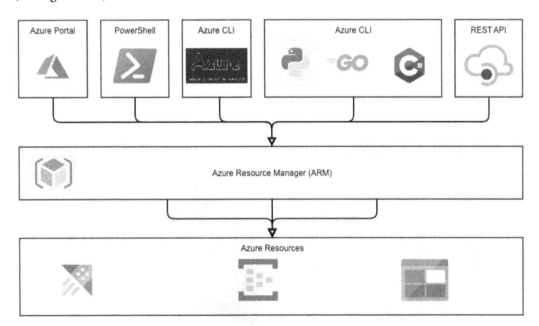

Figure 2.23 – ARM

In the next two sections, *Creating and configuring ADX instances with PowerShell* and *Creating ADX clusters with ARM templates*, we will look at two different programming paradigms for IaC, **imperative** and **declarative programming**.

Imperative programming is a paradigm that uses statements such as `if`, `while`, and `for` to change and manage a program's flow. For example, the PowerShell examples should ideally contain error handling to ensure the previous operations have succeeded. For simplicity, the code examples have no error handling and we assume each cmdlet will be successful, which should never be done in a production environment.

The second paradigm is declarative, which abstracts the need for managing the code's control flow for error handling. For instance, ARM templates are a form of declarative programming. As you will see, with ARM templates, we only need to declare the resources we want to create. The logic and error handling are taken care of by ARM.

Creating and configuring ADX instances with PowerShell

As mentioned earlier, we will use Azure Cloud Shell for this exercise, so if you have been following along, you should have provisioned Azure Cloud Shell. If you have not, please be sure to read the earlier section, *Introducing Azure Cloud Shell*.

Let's begin by cloning the repository in Azure Cloud Shell:

1. Open a new browser tab and go to `https://shell.azure.com`. The URL takes you directly to Cloud Shell.

2. Ensure your current shell is **PowerShell**. You will see the name of your current shell in the top-left corner, as shown in *Figure 2.24*:

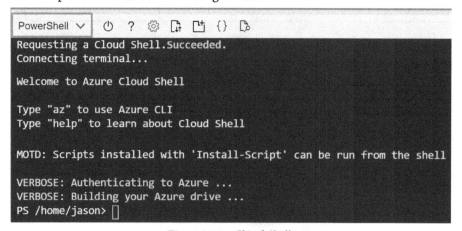

Figure 2.24 – Cloud Shell

3. Create a new directory called `development`, by typing `mkdir development`, and navigate into the directory: `cd development`.

4. Clone our Git repository: `git clone https://github.com/PacktPublishing/Scalable-Data-Analytics-with-Azure-Data-Explorer.git`.

5. Navigate into the `Scalable-Data-Analytics-with-Azure-Data-Explorer` directory by typing `cd Scalable-Data-Analytics-with-Azure-Data-Explorer directory`.

6. And now for one of the best, and not well-known, features of PowerShell. Type `code .` to open a lightweight **Visual Studio Code Editor**, as shown in *Figure 2.25*. The code editor has been a lifesaver in emergency situations where I have had to make code changes or troubleshoot on the spot.

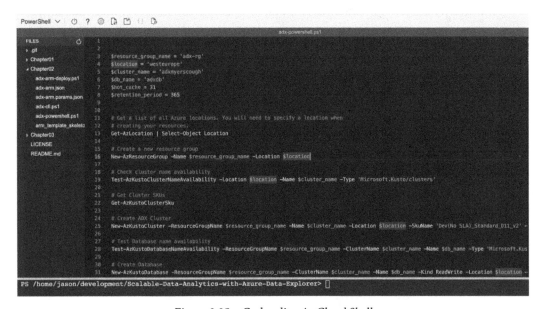

Figure 2.25 – Code editor in Cloud Shell

7. From Cloud Shell, navigate to the `Chapter02` directory by typing `cd Chapter02`.

8. Finally, type `./adx-powershell.ps1`. This will execute the PowerShell script and create our ADX instance. Please keep in mind this deployment can take up to 20 minutes to complete. ADX cluster names must be globally unique. You may get an error during your deployment like the following:

    ```
    New-AzResourceGroupDeployment: 12:22:03 AM - The
    deployment 'AdxDeployment' failed with error(s). Showing
    1 out of 1 error(s).

    Status Message: Name 'adxmyerscough' with type Engine
    is already taken. Please specify a different name
    (Code:InvalidClusterName)
    ```

    If so, update *line 5* (`$cluster_name = 'adxmyerscough'`) and change the name of the cluster from `adxmyerscough` to something unique.

9. To understand what is happening, let's run each command in PowerShell manually:

    - *Lines 3-8* are our variables. The variables are used to improve readability. We have created variables for storing the resource group name, the location to denote the Azure region we want to deploy to, the cluster name, the database name, the hot cache period, and the retention period.

    - *Line 13*, `Get-AzLocation | Select-Object Location`, returns a list of Azure regions. Since we have already initialized `$location` with `westeurope`, we do not technically need this line, but I have added it in case you would like to deploy into a different Azure region.

    - *Line 16*, `New-AzResourceGroup -Name $resource_group_name -Location $location`, creates our resource group.

    - *Line 19*, `Test-AzKustoClusterNameAvailability -Location $location -Name $cluster_name`, checks to ensure the name for the cluster is globally unique. If you recall from earlier, we mentioned that cluster names must be globally unique.

    - *Line 22*, `Get-AzKustoClusterSku`, returns a list of SKU names and tiers for each Azure region. You may recall the Azure portal displayed **Dev** and **Standard** SKU tiers when creating the ADX cluster via the portal. The **Dev** tier is actually named **Basic** and this is the value we must use when creating a cluster via PowerShell.

    - *Line 25*, `New-AzKustoCluster -ResourceGroupName $resource_group_name -Name $cluster_name -Location $location -SkuName 'Dev(No SLA)_Standard_D11_v2' -SkuTier 'Basic' -SkuCapacity 1`, creates our ADX cluster. This operation can take up to 20 minutes to complete.

- *Line 28*, `Test-AzKustoDatabaseNameAvailability -ResourceGroupName $resource_group_name -ClusterName $cluster_name -Name $db_name -Type 'Microsoft.Kusto/ Clusters/Databases'`, checks to ensure the database name is available within the cluster.

- *Line 31*, `New-AzKustoDatabase -ResourceGroupName $resource_ group_name -ClusterName $cluster_name -Name $db_name -Kind ReadWrite -Location $location -HotCachePeriod $hot_cache -SoftDeletePeriod $retention_period`, creates the database in our ADX cluster. Notice how the parameter names, such as `SoftDeletePeriod`, are different from the portal, where it is referred to as the retention period.

We can now destroy and create our ADX cluster and database consistently in a reliable manner by executing our PowerShell scripts using our IaC. In the next section, *Creating ADX clusters with ARM templates*, we will create our cluster again, this time using ARM templates, which is a better way of writing IaC, as we will see next.

Creating ADX clusters with ARM templates

ARM templates are declarative **JavaScript Object Notation** (**JSON**) files that we use to define our infrastructure and configuration requirements. There is a lot of debate surrounding the ease of readability of JSON and at the time of writing, Microsoft has a new tool in preview called **Bicep**, which is similar to **Terraform**'s propriety **HashiCorp Configuration Language** (**HCL**). We are not going to compare ARM with Bicep or Terraform; each tool has a purpose and what you choose ultimately depends on your requirements.

It is not possible to cover all aspects of ARM templates in this short chapter, so we will cover the basics to get started.

ARM template structure

As shown in the following code snippet, ARM templates consist of six sections. There is a seventh section called `functions`, which is rarely used, and I will not cover it here. I have only used the `functions` section once:

```
{
    "$schema": "https://schema.management.azure.com/
schemas/2019-04-01/deploymentTemplate.json#",
    "contentVersion": "1.0.0.0",
```

```
    "parameters": {},
    "variables": {},
    "resources": [],
    "outputs": {}
  }
```

Table 2.2 describes each section of ARM templates:

Section	Required	Data Type	Description
$schema	Yes	string	Specifies the version of the template language. There are two important schemas to be aware of when developing templates. The first is for the templates themselves: `http://schema.management.azure.com/schemas/2019-04-01/developmentTemplate.json#` and the second is for your parameter files: `https://schema.management.azure.com/schemas/2019-04-01/deploymentParameters.json#`
contentVersion	Yes	string	Specifies the version of your template. Even though the contentVersion section is required, I have never used this value for tracking my template version. For versioning and tagging, I use my Git workflow. I normally set this value to 1.0.0.0.
parameters	No	object	Parameters are the key to reusability. Parameters are provided via parameter files that allow us to customize the resource deployment.
variables	No	object	Variables enable us to write expressions and assign the results to variables. A typical use of variables is to append unique identifiers to parameters, concatenate parameters, and pass parameters to functions, which we will see later.
resources	Yes	array	An array of 1 or more Azure resources you wish to create such as Azure Data Explorer instances, virtual machines, virtual networks, and so on.
outputs	No	object	Values that are returned after a deployment. This is useful when you want to reference a resource's property later in the deployment.

Table 2.2 – ARM template sections

Let's look at the parameters, variables, and resources that we have declared in our ARM template: `${HOME}/Scalable-Data-Analytics-with-Azure-Data-Explorer/Chapter02/adx-arm-deploy.ps1`.

Parameters

As mentioned, parameters are at the heart of reusability. They allow us to customize deployments while reusing our template for each deployment. The following code snippet shows how to declare a parameter called `adx_cluster_engineVersion` that is of type `string` and only accepts `V3` as input. The deployment would fail if you tried to specify a different value:

```
"adx_cluster_engineVersion": {
    "type": "string",
    "allowedValues": ["V3"],
    "defaultValue": "V3",
    "metadata": {
        "description": "The Cluster Engine version."
    }
}
```

When declaring parameters, we can specify the data type such as `string`, `int`, and `boolean`. We can also restrict input, for example, specify a max length for a string or define a list of accepted values.

Now let's look at the variables we have declared in our ARM template.

Variables

In our ARM template, we have declared two variables, `adx_cluster_resourceId` and `adx_db_name`.

Our first variable, `adx_cluster_resourceId`, is used to store our cluster's resource ID. Every resource in Azure has a unique identifier. Here we are calling an ARM function called `resourceId()` that returns the ID:

```
"adx_cluster_resourceId": "[resourceId('Microsoft.Kusto/
clusters', parameters('adx_cluster_name'))]",
```

We use the variable on *line 111*, to inform ARM to first deploy the cluster before creating a database:

```
 "dependsOn": [
      "[variables('adx_cluster_resourceId')]"
  ]
```

The second variable concatenates (joins together) the cluster name and database name in the form of `adxmyerscough/adxdemo`. This format is required because the database is a child resource of the ADX cluster:

```
"adx_db_name": "[concat(parameters('adx_cluster_name'), '/',
parameters('adx_database_name'))]"
```

Now let's look at our resources before deploying the template.

Resources

As mentioned earlier, the resources section is an array of objects, with each object representing a resource we want to deploy such as an ADX cluster and ADX database. A complete list of Azure resources and their properties can be found at `https://docs.microsoft.com/en-us/azure/templates/`. Our ARM template consists of two resources, the cluster, and the database. As shown in *Figure 2.26*, *lines 89-104* represent our ADX cluster and *lines 105-117* represent our database:

```
 87    "resources": [
 88        {
 89            "type": "Microsoft.Kusto/clusters",
 90            "apiVersion": "2020-06-14",
 91            "name": "[parameters('adx_cluster_name')]",
 92            "location": "[resourceGroup().location]",
 93            "sku": {
 94                "name": "[parameters('adx_cluster_sku_name')]",
 95                "tier": "[parameters('adx_cluster_sku_tier')]",
 96                "capacity": "[parameters('adx_cluster_sku_capacity')]"
 97            },
 98            "properties": {
 99                "enableDiskEncryption": "[parameters('adx_cluster_enableDiskEncryption')]",
100                "enableDoubleEncryption": "[parameters('adx_cluster_enableDoubleEncryption')]",
101                "enablePurge": "[parameters('adx_cluster_enablePurge')]",
102                "engineType": "[parameters('adx_cluster_engineVersion')]"
103            }
104        },
105        {
106            "type": "Microsoft.Kusto/clusters/databases",
107            "apiVersion": "2020-06-14",
108            "name": "[variables('adx_db_name')]",
109            "location": "[resourceGroup().location]",
110            "dependsOn": [
111                "[variables('adx_cluster_resourceId')]"
112            ],
113            "properties": {
114                "softDeletePeriodInDays": "[parameters('adx_db_softDeletePeriod')]",
115                "hotCachePeriodInDays": "[parameters('adx_db_hotCachePeriod')]"
116            }
117        }
118    ]
```

Figure 2.26 – ARM resources

The following are some of the most common properties of Azure:

- `type`: Specifies the type of Azure resource being deployed, for example, `Microsoft.Kusto/clusters` denotes an ADX cluster and `Microsoft.Kusto/clusters/databases` denotes a database.

- `apiVersion`: Specifies the version of the resource. This is an important property since new versions can introduce new properties to the resource.

- `name`: Specifies the name of the resource. On *line 91*, `"name"`: `"[parameters('adx_cluster_name')]"`, you can see the value is a parameter.

- `location`: Specifies the Azure region where the resource will be deployed. It is best practice to deploy resources in the same location as their resource group. We can use the `resourceGroup()` function to retrieve the location of the resource group. On *lines 92 and 109*, we set our location using the `resourceGroup()` function: `"location"`: `"[resourceGroup().location]"`.

- `properties`: Specifies the resource's properties that we can configure, for example, for the ADX cluster, we can set properties such as `enableDiskEncryption` and `enablePurge`.

A deeper dive into ARM templates could easily take up an entire book. We have barely scratched the surface here, but we now know enough to be able to deploy our templates. In the next section, *Deploying our templates*, we will deploy our ARM template from Cloud Shell.

Deploying our templates

Let's deploy our ADX cluster using our ARM templates. Ideally, we use **Azure DevOps CI/CD pipelines** to deploy our templates, but CI/CD pipelines are beyond the scope of this book, so we are going to execute the PowerShell cmdlets to deploy our ARM templates:

1. Open `${HOME}/Scalable-Data-Analytics-with-Azure-Data-Explorer/Chapter02/adx-arm-deploy.ps1`, copy *line 9*, `New-AzResourceGroup -Name "adx-rg" -Location "westeurope"`, and paste the cmdlet into Cloud Shell and hit *Enter*. This command will create a new resource group called **adx-rg** in West Europe. Like the PowerShell example, you may get an error such as the following:

```
New-AzResourceGroupDeployment: /Users/JasonM/development/
Scalable-Data-Analytics-with-Azure-Data-Explorer/
Chapter02/adx-arm-deploy.ps1:12:1
Line |
   12 |   New-AzResourceGroupDeployment -Name
"AdxDeployment" -ResourceGroupNam …
      |   ~~~~~~~~~~~~~~~~~~~~~~~~~
```

```
| 15:14:09 - The deployment 'AdxDeployment' failed
with error(s). Showing 1 out of 1 error(s). Status
Message: Name 'adxmyerscough' with type Engine is already
taken. Please specify a different name

| (Code:InvalidClusterName)  CorrelationId:
5f41cc32-3536-472e-8d21-37a2557ef59e
```

If so, update the parameter file, Chapter02/adx-arm.params.json, and update the value of adx_cluster_name on *line 6* to something that is globally unique.

2. Copy line 12, New-AzResourceGroupDeployment -Name "AdxDeployment" -ResourceGroupName "adx-rg" -TemplateFile ./adx-arm.json -TemplateParameterFile ./adx-arm.params.json, paste it into Cloud Shell, and hit *Enter*. This command will deploy your ARM template along with the parameter file: adx-arm.params.json. This operation can take up to 20 minutes to complete.

Cloud Shell will output a deployment summary, as shown in *Figure 2.27*, once the deployment has successfully finished:

```
PS /home/jason/development/Scalable-Data-Analytics-with-Azure-Data-Explorer> cd ./Chapter02/
PS /home/jason/development/Scalable-Data-Analytics-with-Azure-Data-Explorer/Chapter02> New-AzResourceGroupDeployment -N
lateParameterFile ./adx-arm.params.json

DeploymentName          : AdxDeployment
ResourceGroupName       : adx-rg
ProvisioningState       : Succeeded
Timestamp               : 6/12/2021 1:23:20 PM
Mode                    : Incremental
TemplateLink            :
Parameters              :
                          Name                                Type                        Value
                          ================================    ======================      ==========
                          adx_cluster_name                    String                      adxmyerscough
                          adx_database_name                   String                      adx-db
                          adx_cluster_sku_name                String                      Dev(No SLA)_Standard_D11_v2
                          adx_cluster_sku_tier                String                      Basic
                          adx_cluster_sku_capacity            Int                         1
                          adx_cluster_enableDiskEncryption    Bool                        True
                          adx_cluster_enablePurge             Bool                        False
                          adx_cluster_enableDoubleEncryption  Bool                        False
                          adx_cluster_engineVersion           String                      V3
                          adx_db_softDeletePeriod             Int                         365
                          adx_db_hotCachePeriod               Int                         30

Outputs                 :
DeploymentDebugLogLevel :
```

Figure 2.27 – ARM template deployment summary

The beauty of ARM templates is their reusability. If we wanted to deploy another ADX cluster, we would only need to create a new parameter file with the cluster details and then deploy the cluster using the two aforementioned PowerShell cmdlets.

Summary

Well done! You made it this far. We covered a lot of topics, and you should be proud of what we accomplished. We started by creating our first *subscription* and activated our free trial, which is valid for 30 days at the time of writing. Then we learned about Cloud Shell, which is a web-based console that provides *PowerShell* and *Bash* terminals directly in the Azure portal. We also learned about the lightweight code editor embedded in *Cloud Shell*, which is a very convenient feature. We then provisioned our own Cloud Shell instance, allowing us to create *ADX clusters and databases* via PowerShell and *ARM templates*.

Then, we created our first ADX cluster and database using the Azure portal, then deleted it and learned about the benefits of *Infrastructure as Code* and introduced the declarative and imperative programming paradigms.

Then, we learned how to create ADX clusters and databases using PowerShell cmdlets and had a quick introduction to ARM templates, where we learned about the structure of templates, how to use parameter files to customize our deployments, and finally deployed our ARM template to create a new ADX instance.

In the next chapter, we will learn about the **ADX management interface**, where we will spend most of our time when we start querying data and generating visuals.

Questions

Before moving on to the next chapter, test your knowledge by trying these exercises. The answers can be found at the back of the book:

1. Modify `adx-powershell.ps1` and try to deploy another cluster with `doubleEncryption` enabled. Hint: `New-AzKustoCluster` has an optional parameter called `-EnableDoubleEncryption`.

2. Create a new parameter file and enable purging and `doubleEncryption`, then change the *hot cache period to 10 days* and the *soft delete period to 100 days*.

3. In the Azure portal, create a second ADX database and set the hot cache to 10 days and soft delete to 50 days.

4. Modify `adx-powershell.ps1` and deploy your ADX cluster to an Azure region that is close to you.

5. What is the difference between the hot cache and soft delete retention periods?

6. What shells are supported by Cloud Shell?

7. How do you open the code editor in Cloud Shell?

3
Exploring the Azure Data Explorer UI

In *Chapter 1*, *Introducing Azure Data Explorer*, we introduced the **Azure Data Explorer (ADX) Web UI** and executed our first query, but we did not explore the various options and panels in detail. In the previous chapter, we focused on configuring and deploying our **ADX clusters and databases**, but we have not yet started to ingest data to query. In this chapter, we will cover the ADX Web UI in detail and introduce you to data ingestion using a method called **one-click ingestion**.

Although it is technically possible to run queries and generate visualizations from within the **Azure portal**, which we will demonstrate in this chapter, the Azure portal can feel distracting and cluttered with all the additional, non-ADX-related menus and options such as creating resources, Cloud Shell, and so forth. The ADX UI is a clean, uncluttered **user interface** (UI) that allows you to focus on exploring your data and provides options for ingesting data and creating dashboards that are not available in the Azure portal's embedded UI.

In this chapter, we will begin by ingesting a sample dataset from Microsoft so that you can follow along and gain experience using the ADX UI. We will then learn how to query our databases directly in the Azure portal and understand some of the limitations.

Next, we will focus on the ADX UI, learn about how to navigate the UI, and understand the different options and features.

> **Note**
> This chapter is intended to be an introduction to the features of the ADX Web UI. Most of the features discussed will be used in later chapters. This chapter can be used as a quick reference guide as you work through the later chapters. Microsoft also provides a desktop rich client called Kusto Explorer. More information and the installer can be found at: `https://aka.ms/ke`.

In this chapter, we are going to cover the following main topics:

- Ingesting the `StormEvents` sample dataset
- Querying data in the Azure portal
- Exploring the ADX Web UI

Technical requirements

All the examples in this chapter are carried out via the Azure portal and the ADX Web UI. The example queries used can be found at `https://github.com/PacktPublishing/Scalable-Data-Analytics-with-Azure-Data-Explorer` in the `Chapter03` directory.

Ingesting the StormEvents sample dataset

In the spirit of keeping the chapter both theoretical and practical, we are going to jump ahead a little and ingest an example dataset that Microsoft provides. Don't worry if you do not understand all the details, as data ingestion will be discussed in *Chapter 4, Ingesting Data in Azure Data Explorer*.

In the previous chapter, we created our ADX cluster and databases, but we did not ingest any data or create any tables. We are going to use a method called **one-click ingestion**, which is amazingly simple to use and is a great example of ADX allowing you to focus on exploring your data rather than worrying about the low-level details of ingestion.

As you will recall from *Chapter 2, Building Your Azure Data Explorer Environment,* the *third step* in the creation process is to ingest data, as shown in the following screenshot:

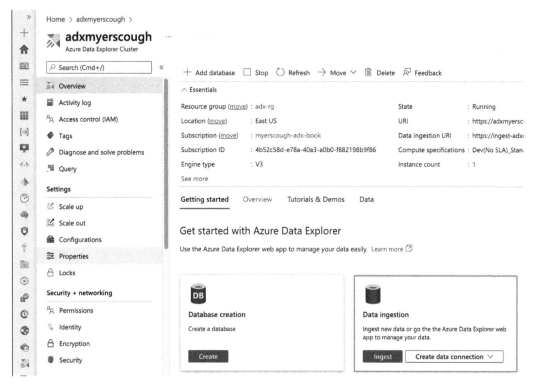

Figure 3.1 – Data ingestion

The following sequence of steps will import Microsoft's sample dataset into our ADX database:

1. Open the Azure portal by going to `https://portal.azure.com` and click on your ADX cluster. The cluster will be listed under **Recent resources**, as shown in the following screenshot:

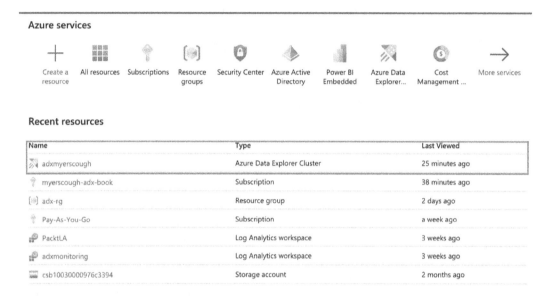

Azure services

| Create a resource | All resources | Subscriptions | Resource groups | Security Center | Azure Active Directory | Power BI Embedded | Azure Data Explorer... | Cost Management ... | More services |

Recent resources

Name	Type	Last Viewed
adxmyerscough	Azure Data Explorer Cluster	25 minutes ago
myerscough-adx-book	Subscription	38 minutes ago
adx-rg	Resource group	2 days ago
Pay-As-You-Go	Subscription	a week ago
PacktLA	Log Analytics workspace	3 weeks ago
adxmonitoring	Log Analytics workspace	3 weeks ago
csb10030000976c3394	Storage account	2 months ago

Figure 3.2 – Recent Azure resources

2. Click **Ingest new data**, as shown in *Figure 3.1*. A new browser tab will open and take you to the ADX Web UI's **Ingest new data** page, as shown in the following screenshot:

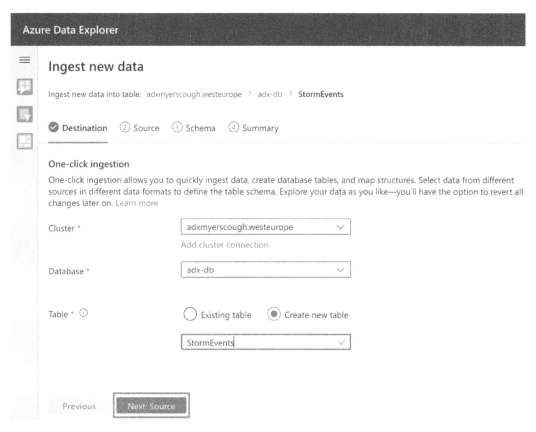

Figure 3.3 – Configuring data ingestion

3. Select the cluster and database that you created in the previous chapter.

4. Create a new table by clicking **Create new**, name it `StormEvents`, and click **Next: Source**.

5. Select **From Blob** as the source type. **From Blob** refers to **Blob storage**. Blob storage is Azure's fundamental storage service that is highly scalable, allowing you to store large amounts of data. Other Azure storage services such as **file shares** and **Data Lake** are built on top of Blob storage.

6. Use `https://kustosamplefiles.blob.core.windows.net/samplefiles/StormEvents.csv?sv=2019-12-12&ss=b&srt=o&sp=r&se=2022-09-05T02:23:52Z&st=2020-09-04T18:23:52Z&spr=https&sig=VrOfQMT1gUrHltJ8uhjYcCequEcfhjyyMX%2FSc3xsCy4%3D` to populate the **Link to source** field.

7. Next, click **Next: Schema** to navigate to the **Schema** blade. As shown in the following screenshot, on the left you can see that ADX has correctly determined the compression type and file type and has recommended skipping the first record, which is the column headers:

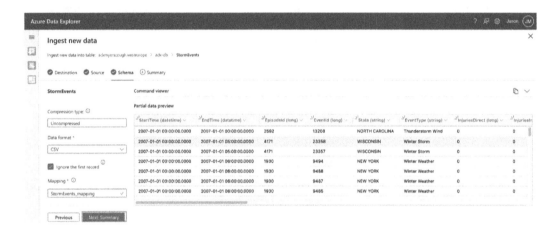

Figure 3.4 – Schema configuration

8. Click **Close** once the data ingestion process is complete. The amount of time required to complete the ingestion can vary depending on how busy the Azure platform is.

9. Click **Next: Summary** to ingest the data. As shown in *Figure 3.5*, a summary will be displayed verifying the data has been ingested:

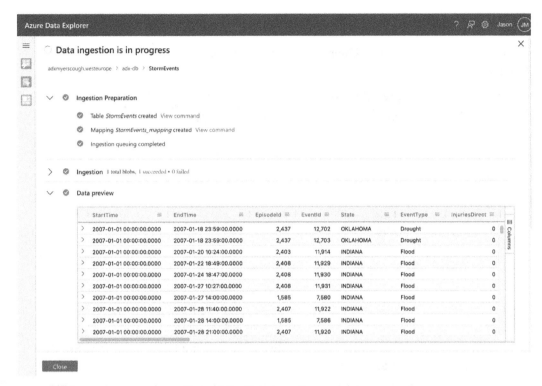

Figure 3.5 – Data ingestion summary

Although the **one-click ingestion** process is straightforward and robust, this can at times fail. At the time of writing, I experienced a couple of instances where the **Ingestion Preparation** phase failed when trying to queue the ingestion request. The `StormEvents` table was created, but there was no data. The **Ingestion Preparation** phase consists of three steps:

1. Creating a database table.
2. Creating a mapping schema.
3. Queueing the ingestion request.

If you get an error such as `Failed to queue blob for ingestion. Error:'Failed to send request to https://z5mkstrldadxmyerscough00.queue.core.windows.net/aggregatorinput-secured-3/messages?sv=2018-03-28&sig=IIc6r3lqJYwGeFQQQ8AeGqZhygeZ8gO85FR7yIMsYQs%3D&st=2021-08-07T02%3A50%3A50Z&se=2021-08-07T14%3A55%3A50Z&sp=a&timeout=30',` my recommendation is to drop the table and repeat the ingestion process.

The following sequence of steps explains how to repeat the ingestion process:

1. In the ADX Web UI, click **Query**, as shown in *Figure 3.6*. You will be taken to the query editor:

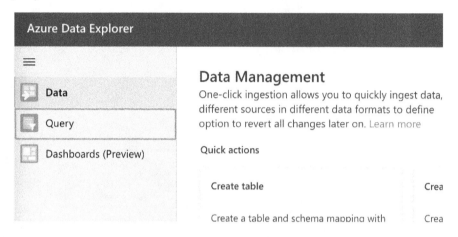

Figure 3.6 – ADX Web UI's navigation panel

2. Next, right-click the StormEvents table and click **Drop table** to delete it. See the following screenshot for an overview of this:

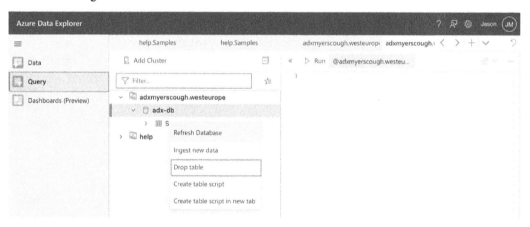

Figure 3.7 – Deleting the StormEvents table

3. Finally, click **Data** in the navigation panel and select **Ingest new data** from the **Quick actions** list. Now, repeat the steps mentioned earlier to ingest your data.

Now that we have ingested the sample dataset, let's see how we can query the data using the Azure portal.

Querying data in the Azure portal

The ADX UI can be embedded in any web page as a **HyperText Markup Language (HTML) inline frame (iFrame)** and, as shown in *Figure 3.8*, this is how the UI is integrated into the Azure portal. The embedded UI is a lightweight version of the ADX UI that includes the **Cluster**, **Query Editing**, and **Query Results** panels. You cannot create ADX dashboards or ingest data from within the Azure portal, and if you try ingesting data by right-clicking your cluster, you will be redirected to the full web UI (`https://dataexplorer.azure.com/`). Since the embedded UI is a lightweight version of the ADX Web UI, we will discuss the **Cluster**, **Query Editing**, and **Query Results** panels in the following section, *Exploring the ADX Web UI*:

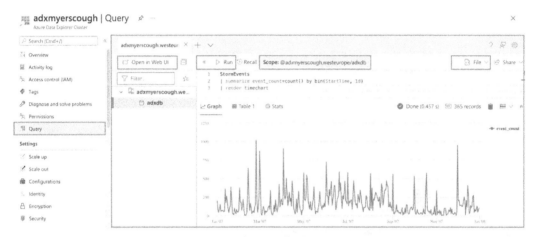

Figure 3.8 – Querying data from the Azure portal

> **Note**
>
> As mentioned, the ADX UI can be embedded in any web page as an HTML iFrame. If you are interested in embedding the UI in your own web page, then I recommend you read the following: `https://docs.microsoft.com/en-us/azure/data-explorer/kusto/api/monaco/monaco-kusto`.

The following sequence of steps demonstrates how to query your tables in the Azure portal using the lightweight **Query Editor:**

1. Open the Azure portal (`https://portal.azure.com`) and click on your ADX cluster.

2. Under your ADX cluster's properties, click **Query**. This will open the embedded UI in a new blade, as shown in *Figure 3.8.*

3. From the UI blade, click **File | Open**, and open `${HOME}/Scalable-Data-Analytics-with-Azure-Data-Explorer-/Chapter03/dailyevents.kql`.

4. Click **Run**. As shown in *Figure 3.8*, a time chart will be rendered showing the number of daily events.

> **Note**
>
> When executing a query, ensure you are at the correct scope—in our case, the database scope. You can do this by expanding the cluster and selecting the database. As shown in *Figure 3.8*, the scope is set to `@adxmyerscough.westeurope/adxdb`. If you are not at the correct scope, the query will not return any results.

Now that we have seen how to query data in the Azure portal and learned how the Azure portal uses the embedded UI, we will look at the different panels and options in the ADX Web UI in the next section, *Exploring the ADX Web UI*. The remainder of the book will use the ADX Web UI when querying data.

Exploring the ADX Web UI

By now, you will be aware that the ADX Web UI can be accessed via `https://dataexplorer.azure.com/`. Use the same credentials you use to log in to the Azure portal. Once you are logged in, you will see the ADX Web UI, as shown in *Figure 3.9*:

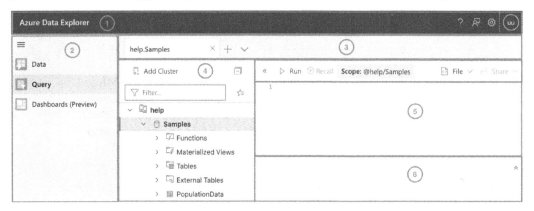

Figure 3.9 – ADX Web UI

Each of the panels shown in *Figure 3.9* is described in more detail here:

1. The **header** is where you can find access to help documentation, your user account, and some UI configuration settings.

2. In the **Navigation** panel, unlike the embedded UI in the Azure portal, the Web UI provides two more views: **Data** and **Dashboards**. As shown in the following screenshot, the **Data** view allows you to configure data ingestion using the **one-click ingestion** method, which we used earlier in the chapter to ingest the sample dataset. We will cover data ingestion in more detail in the next chapter:

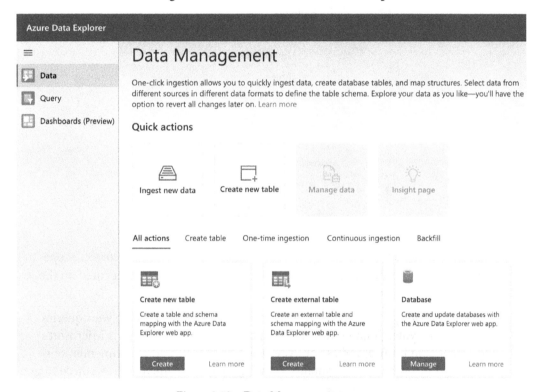

Figure 3.10 – Data Management screen

The **Query** view is where we can write and execute our queries and view the results. The **Query** view is the same as the embedded UI view in the Azure portal.

The **Dashboard** view is currently in preview and is the view where we can create dashboards to share the results of our data analysis with key stakeholders. We will look at this more in *Chapter 8, Data Visualization with Azure Data Explorer and Power BI*.

3. The tab bar allows us to have multiple queries and query editors open.

4. The **cluster connection** panel is where we can manage our clusters. As shown in the following screenshot, it is possible to connect to multiple ADX clusters:

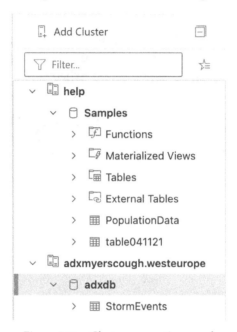

Figure 3.11 – Cluster connection panel

Earlier in the chapter, I mentioned that it is important to ensure the correct scope is selected when executing queries. The scope ensures the query is executed on the correct cluster and database.

5. The **Query Editor** panel is the view where you can write and execute your queries. The editor has syntax highlighting and supports **IntelliSense**, which is Microsoft's code completion feature, and as you type, it provides hints and recommendations. The **Query Editor** panel also contains a simple menu bar, which is worth discussing. As shown in *Figure 3.12*, the menu consists of two buttons—a view of the current scope and a **File** and **Share** menu:

Figure 3.12 – Query Editor menu

The **Run** and **Recall** buttons are used to execute your queries. **Run** executes your query and **Recall** returns the result set of the last query without executing the query a second time. The **Scope** field shows where the query will be executed, and the **File** menu can be used to open and save queries and to export query results to a **comma-separated values (CSV)** file. The **Share** menu allows us to share queries and pin them to dashboards.

6. The **Query Results** panel is where the results of our queries will be displayed. The results can be disabled in either tabular or graphical form. Another important view on the **Query Results** panel is the **Stats** view. As shown in *Figure 3.13*, the **Stats** view displays performance telemetry, such as resource consumption and cache performance, which is useful for troubleshooting and performance tuning:

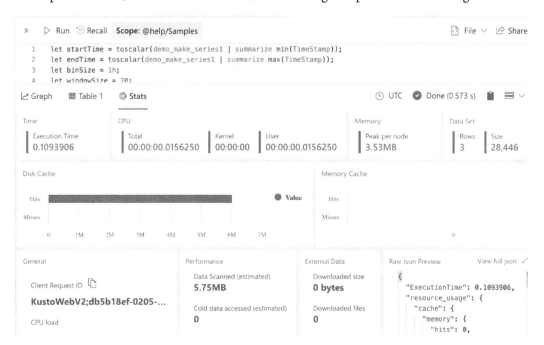

Figure 3.13 – ADX query statistics

All the features mentioned here will be used throughout the remainder of the book. Unless we are modifying the properties of our ADX cluster, most of our time will be spent in the ADX Web UI.

Summary

There you go! That was our introduction to the main features of the ADX UI. In this chapter, we ingested Microsoft's sample dataset using the **one-click ingestion** method and we learned how to query data from within the Azure portal.

The remainder of the chapter looked at the main panels and features of the ADX Web UI. We saw how the Azure portal uses the embedded UI and that the embedded UI can be embedded in any web page using **iFrames**. Next, we learned that the Web UI (`https://dataexplorer.azure.com/`) consists of three main views/windows.

The **Data** view allows us to ingest data using the **one-click ingestion** method.

The **Query Editor** view allows us to write and execute our queries and provides syntax highlighting and Microsoft's IntelliSense.

The **Dashboards** view allows us to create dashboards based on our queries that we can share with our stakeholders. We will cover dashboards in more detail in *Chapter 8, Data Visualization with Azure Data Explorer and Power BI*.

We also reviewed the **Stats** tab on the **Query Results** view, which provides valuable insights into query performance.

In the next chapter, we will discuss data ingestion in more detail and look at the different methods of ingesting data.

Section 2: Querying and Visualizing Your Data

This section of the book focuses on data ingestion and how to query and visualize your data using the powerful KQL query language. It begins by introducing you to data ingestion, describing the different types and sources of data that can be ingested. It then introduces you to the powerful, read-only querying language: KQL. *Chapter 5, Introducing the Kusto Query Language*, covers most of the KQL operators and functions required to efficiently query and explore your data. Finally, this section discusses how to perform time series analysis and how to search for anomalies and trends in your data.

This section consists of the following chapters:

- *Chapter 4, Ingesting Data in Azure Data Explorer*
- *Chapter 5, Introducing the Kusto Query Language*
- *Chapter 6, Introducing Time Series Analysis*
- *Chapter 7, Identifying Patterns, Anomalies, and Trends in Your Data*
- *Chapter 8, Data Visualization with Azure Data Explorer and Power BI*

4
Ingesting Data in Azure Data Explorer

In the previous chapters, we created our **Azure Data Explorer** (**ADX**) clusters and databases, learned how to use the **Data Explorer Web** UI, executed our first **Kusto Query Language** (**KQL**) query. Now, we are ready to look at data ingestion in detail and start to ingest data. **Data ingestion** is the process of taking data from an external source and importing it into your big data solution, in our case, ADX. As you will soon see, once the data has been ingested, we can begin to analyze the data and generate visuals such as graphs and reports.

In this chapter, we will introduce data ingestion and discuss the different types of data (structured, semi-structured, and unstructured). Then we will examine the different data ingestion methods that ADX supports and learn how ADX ingests data via its **Data management service**, which we introduced in *Chapter 1, Introducing Azure Data Explorer*.

Next, we will learn about **schema mapping**, which is the process of mapping external data to columns in our Azure Data Explorer database tables. We will learn how to write **ordinal-** and **path-based** schema maps for tabular and structured formats. We will be using a dataset that contains the football results for the English Premier League (EPL) for the last 10 years.

Then we will learn how to ingest the dataset using a variety of methods. We will begin by using the **one-click ingestion** method in the Data Explorer Web UI, then we will ingest the data using **Kusto Query Language** (**KQL**) management commands, and finally, we will ingest our datasets from Blob storage, which is Azure's object storage service using Azure Event Grid.

In this chapter, we are going to cover the following main topics:

- Understanding data ingestion
- Introducing schema mapping
- Ingesting data using one-click ingestion
- Ingesting data using KQL management commands
- Ingesting data from Blob storage using Azure Event Grid

Technical requirements

The code examples for this chapter can be found in the `Chapter04` folder of the repo: `https://github.com/PacktPublishing/Scalable-Data-Analytics-with-Azure-Data-Explorer.git`. The `Chapter04` directory contains two directories, `templates/`, which contains our ARM templates, and `datasets/`, which contains our datasets that we will be ingesting.

One of the challenges when it comes to writing about data analytics is to have interesting datasets that are large enough to demonstrate the features of ADX and KQL. In this chapter, we will use the English Premier League's results to demonstrate how to ingest data in CSV and JSON format. A copy of the data is included in our repository and the original dataset can be found at `https://datahub.io/sports-data/english-premier-league`. The dataset provides Premier League results for the last 10 years.

> **Note**
>
> The infrastructure that we will deploy here will be reused later in the book. Feel free to either preserve the resources for later use or delete them once you complete this chapter. You can re-deploy them again when they are needed in *Chapter 9, Monitoring and Troubleshooting Azure Data Explorer.*

Understanding data ingestion

Before learning how data ingestion works with ADX, let's revisit the different types of data:

- **Structured data**: When we think of structured data, we think of relational databases that are made up of tables consisting of rows and columns. Each column has a data type such as an integer or string, and it sometimes includes additional constraints such as fixed-length strings and strings with specific formats such as a postcode.

- **Semi-structured**: When we think of semi-structured data, we think of JSON and XML. They have a structure defined with tags, but the format is typically less rigid than relational databases.

- **Unstructured data**: Unstructured data is data that has no constraints, such as SMS messages, text files, and emails, and social media such as status posts, messages, and images.

As shown in *Figure 4.1*, ADX supports four categories of services that enable data ingestion:

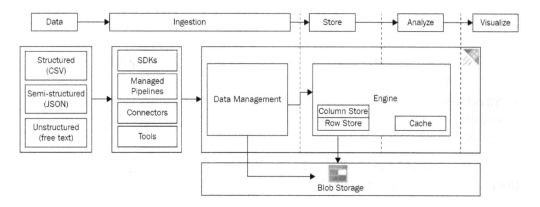

Figure 4.1 – Data analysis pipeline

The four categories of data ingested are listed and described here:

- **SDKs**: Microsoft provides SDKs in various programming languages such as **Go**, **Python**, and **.NET Core** that you can use to implement ingestion straight from your applications.

- **Managed Pipelines**: Managed pipelines are Azure services such as **Event Hubs**, **Event Grid**, and **IoT Hub**. In this chapter, we will learn how to ingest data from **storage accounts** using **Event Grid** and **Event Hubs**.

- **Connectors and Plugins**: These allow us to integrate with third-party products such as **Kafka** and **Apache Spark** with our Azure Data Explorer instance.

- **Tools**: Microsoft provides a couple of convenience tools such as **one-click ingestion** and **LightIngest,** which is a command-line tool, and **Azure Data Factory**, which is a data transformation service. Later in this chapter, we will learn how to use **one-click ingestion**.

As we discussed in *Chapter 1*, *Introducing Azure Data Explorer*, ADX consists of two primary services, the **Engine**, and the **Data management** service. The engine is responsible for processing ingested data, query execution, and so on. The data management service is responsible for several tasks, including managing data ingestion.

ADX pulls the external data that is pushed via SDKs, managed pipelines, connectors and plugins, and tools. The data is sent to the data management service, which ensures the data is ingested correctly, performs error handling, uses the mapping schemas to ensure the data is mapped correctly to the tables, and compresses the data to improve performance. Finally, the Data management service persists the data to storage and makes the table available for querying.

As we know, ADX supports two forms of data ingestion – batch and streaming:

- **Batch ingestion**: Batch ingestion is the preferred ingestion method and is designed for high ingestion throughput. As we will see in *Chapter 11*, *Performance Tuning in Azure Data Explorer*, we can tune batch ingestion by modifying the batch processing policy.
- **Streaming ingestion**: Streaming allows near real-time ingestion for small amounts of data. As mentioned earlier, data is initially ingested to the row store and then moved to the column store.

Now that we understand how ADX ingests data, let's look at schema mapping in the next section, *Introducing schema mapping*, before we start ingesting data.

Introducing schema mapping

As you know, before we can ingest any data into our ADX instance, we need to create tables in our database to store the data. Similar to a SQL database, ADX tables are two-dimensional, meaning they consist of rows and columns. When we create tables, we need to declare the column name and data type. A data type refers to the type of data a column can store, such as strings, dates, and numbers. We will create our own tables later in the chapter.

How do we ensure the data we are ingesting is imported into the correct tables and rows? The destination table is specified during the data connection creation and the columns are mapped to the incoming data using schema maps.

As you will see in the section *Ingesting data from Blob storage using Azure Event Grid*, it is possible for the source file to contain more columns than you are interested in. We will take a data source with over 60 columns and create a schema map to ingest only the columns we are interested in.

The two mapping schemas we are going to look at in this book are **CSV ordinal-based schemas** and **path-based JSON schemas**. Ordinal schemas essentially index the columns in the source data, similar to how you index an array; likewise, the index starts from 0.

For example, *Figure 4.2* shows a simple data source with two columns, name and dob:

name [0]	dob [1]
Harrison	17/04/2012
James	09/07/2014
Sofia	07/03/2016
Lukas	16/03/2018

Figure 4.2 – Data source with two columns

The following schema maps the ADX table columns, name and dob to the ordinal, positional columns of our data source:

```
[
    {"Column": "name", "Datatype": "string", "Properties":
{"Ordinal": "0"}},
    {"Column": "dob", "Datatype": "string", "Properties":
{"Ordinal": "1"}},
]
```

JSON schema maps are path-based. For example, the following JSON data source is mapped to an ADX table with two columns – name and dob:

```
{
    "name": "Harrison",
    "dob": "17/04/2012",
    "nickname": "cato"
}
{
    "name": "James",
    "dob": "09/07/2014"
    "nickname": "bearo"
}
```

The previous JSON example contains three columns, but we are only interested in the first two, name and dob. The following example is a JSON mapping. As you can see, the mapping is similar to the ordinal mapping, with regard to syntax. $. refers to the top/root of the document. Since we are not mapping the third column, nickname, we can ignore this column:

```
[
    {"column": "name", "Datatype":"string", "Properties":
{"Path": "$.Name"}},
    {"column": "dob", "Datatype":"string", "Properties": {"Path":
"$.Dob"}},
]
```

If the table already exists, then we can omit Datatype from the mapping. I have added it for completeness, but since we always create our tables first, the schemas we create will not contain Datatype.

For reference purposes, *Figure 4.3* lists the supported data types:

Type	.NET Type	Description
Bool	System.Boolean	Stores Boolean values - true and false
Datetime	System.DateTime	Stores date and time values
Dynamic	System.Object	Can be used to store structured and semi-structured data
Int	System.Int32	Stores whole numbers
Long	System.Int64	Stores whole numbers
real,double	System.Double	Stores numbers with decimal places
String	System.String	Stores strings of text
timespan,time	System.TimeSpan	Stores time difference or duration
Decimal	System.Data.SqlTypes.SqlDecimal	Stores numbers with decimal places, such as 1.1234

Figure 4.3 – Supported data types

We will create our own mapping schemas later in the book. In the next section, *Ingesting data using one-click ingestion*, we will begin ingesting data into our ADX database.

Ingesting data using one-click ingestion

In this section, we are going to learn how to ingest data using the Data Explorer Web UI and from an Azure Storage account using the one-click ingestion method.

> **Note**
>
> If you have not already cloned the Git repository, please do so now, so you can follow the example. The repository can be found here: `https://github.com/PacktPublishing/Scalable-Data-Analytics-with-Azure-Data-Explorer.git`.

Data ingestion is a three-step process:

1. **Ingestion Preparation**: During the preparation phase, the table and mapping schemas are created.

2. **Ingestion**: The file is then pulled from the queue, which is temporarily stored on an internal storage account, `https://9qwkstr1dmyerscoughadx01.blob.core.windows.net/20210614-ingestdata-e5c334ee145d4b43a3a2d3a96fbac1df-0/1623671437639_season-1819_csv.csv`, and then ingested.

3. **Data Preview**: Once the data has been ingested, it can be previewed and is ready for you to begin querying.

The following section describes how to use the **one-click ingestion** method in the Azure Data Explorer Web UI:

1. Log in to the Data Explorer Web UI by going to `https://dataexplorer.azure.com/`.

2. Click **Add Cluster** and enter the URL of your **ADX** cluster (for instance, my ADX instance is called `https://myerscoughadx.westeurope.kusto.windows.net`), and click **Add** to connect to the instance.

3. From the left-hand side menu, click **Data**, as shown in *Figure 4.4*. Ensure you have expanded your cluster and have clicked your database. This is to ensure you are at the correct scope level, otherwise, the one-click ingestion method will not work. As shown in *Figure 4.4*, my scope is @myerscoughadx.westeurope/adxdemo:

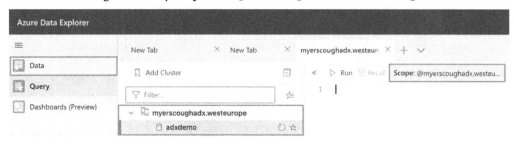

Figure 4.4 – Data management menu

4. As shown in *Figure 4.5*, under **Ingest new data**, click **Ingest**:

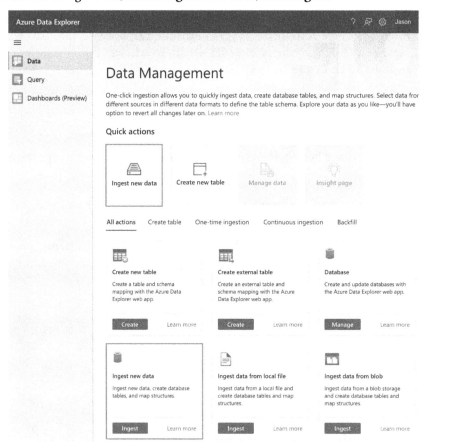

Figure 4.5 – Ingest new data using one-click ingestion

5. For **Cluster**, select our ADX cluster.

6. For **Database**, select our adxdemo database.

7. For **Table**, click **Create new**, create a new table called EPL, and click **Next: Source**.

8. For **Source type**, select from **File**.

9. Drag your file from your local machine into the web browser.

   For example, I uploaded the following:

   ```
   ${HOME}/Scalable-Data-Analytics-with-Azure-Data-Explorer/
   Chapter04/datasets/premierleague/csv/season-1819_csv.csv
   ```

10. Click **Next: Schema**. As shown in *Figure 4.6*, this shows a preview of the data, determines the data types for our columns, creates a schema mapping, and infers the file format. There is also a command viewer that, when expanded, shows the KQL management commands used to create the table and the mapping schema.

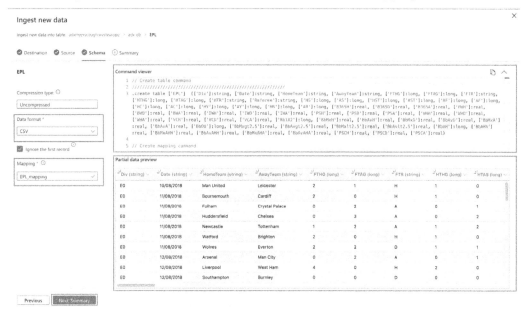

Figure 4.6 – Data ingestion preview

11. Click **Next: Summary** to begin ingesting the data.

12. As shown in *Figure 4.7*, click **Close** and then click **Query menu** in the left-hand menu:

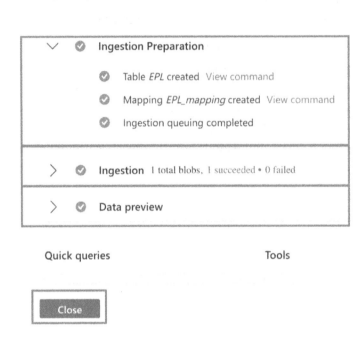

Figure 4.7 – Ingestion summary

13. Enter the following query in the query editor and click **Run**:

```
EPL
| limit 10
```

As shown in *Figure 4.8*, the query displays 10 results from our newly created English Premier League (EPL) table:

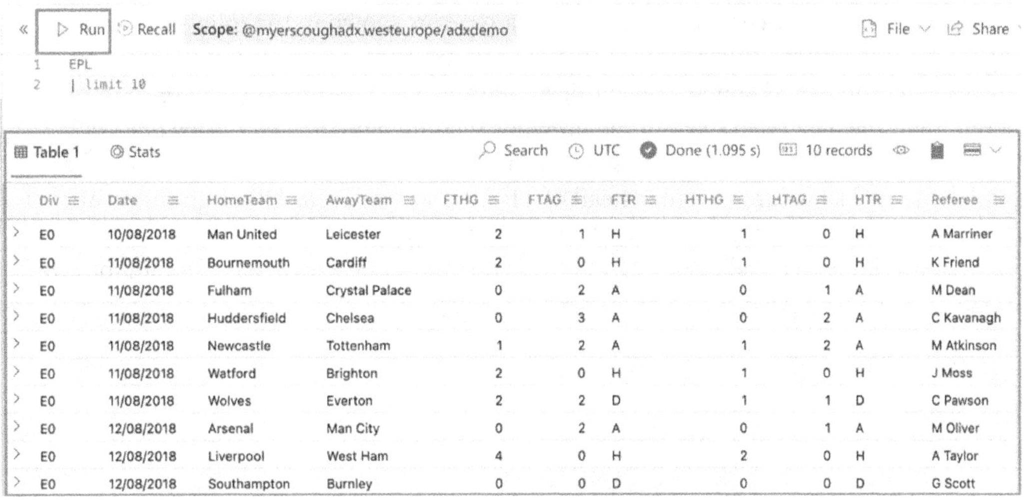

Div	Date	HomeTeam	AwayTeam	FTHG	FTAG	FTR	HTHG	HTAG	HTR	Referee
E0	10/08/2018	Man United	Leicester	2	1	H	1	0	H	A Marriner
E0	11/08/2018	Bournemouth	Cardiff	2	0	H	1	0	H	K Friend
E0	11/08/2018	Fulham	Crystal Palace	0	2	A	0	1	A	M Dean
E0	11/08/2018	Huddersfield	Chelsea	0	3	A	0	2	A	C Kavanagh
E0	11/08/2018	Newcastle	Tottenham	1	2	A	1	2	A	M Atkinson
E0	11/08/2018	Watford	Brighton	2	0	H	1	0	H	J Moss
E0	11/08/2018	Wolves	Everton	2	2	D	1	1	D	C Pawson
E0	12/08/2018	Arsenal	Man City	0	2	A	0	1	A	M Oliver
E0	12/08/2018	Liverpool	West Ham	4	0	H	2	0	H	A Taylor
E0	12/08/2018	Southampton	Burnley	0	0	D	0	0	D	G Scott

Figure 4.8 – Querying our EPL data

In this section, we learned how to ingest data using the one-click ingestion method by uploading one of our data files via the Data Explorer Web UI. You may have noticed the dataset contains some columns related to betting that we are not interested in. In the next section, *Ingesting data using KQL management commands*, we will create our own table and custom mapping schema to ignore the columns we are not interested in.

Ingesting data using KQL management commands

In the previous section, we imported our English Premier League data and you may have noticed that over half of the columns were related to betting statistics. In this section, we will create a custom CSV mapping schema and exclude those columns.

We will also introduce some *KQL* management commands. Like **SQL**, KQL has two categories of commands – data and management. The data commands allow us to query our data and the management commands allow us to manage our clusters, databases, tables, and schemas. We will cover KQL in depth in the next chapter, *Introducing Kusto Query Language*.

The first step is to create a table with the columns that we are interested in. When creating tables, we use the `.create table` command.

We will now specify our columns and their data types as shown in the following code snippet. Here, we are creating a table with clear column names and are not including any of the betting statistics. You may have noticed that the data types used here are slightly different than the data types used in the previous section, *Ingesting data using one-click ingestion*. In the previous section, the schema mapping was generated automatically for us and contained all columns, and instead of `int`, `long` was used. Although this is not technically wrong, `long` types are 64 bit, whereas `int` types are 32-bit. 32-bit integers can hold a maximum integer value of 2,147,483,647, which is more than enough for our football statistics, so I have used `int` instead of `long`:

```
.create table EnglishPremierLeague (
    Divison: string,
    Date: datetime,
    HomeTeam: string,
    AwayTeam: string,
    ...
)
```

The complete source code can be found at `${HOME}/Scalable-Data-Analytics-with-Azure-Data-Explorer/Chapter04/createEplTable.kql`.

Now that we have a table, we need to create a mapping schema to map the external data to columns in our database tables. To do this, we use the `.create table` command with some extra parameters, as shown in the following code snippet.

Here, we are creating a CSV mapping, `.create table ['EnglishPremierLeague'] ingestion csv mapping 'EPL_Custom_Mapping'`, called `EPL_Custom_Mapping`. The remaining lines simply map our database table's columns to the fields in the CSV by referencing their ordinal position:

```
.create table ['EnglishPremierLeague'] ingestion csv mapping
'EPL_Custom_Mapping'
'['
    '{"column": "Divison", "Properties": {"Ordinal": "0"}},'
    '{"column": "Date", "Properties": {"Ordinal": "1"}},'
    '{"column": "HomeTeam", "Properties": {"Ordinal": "2"}},
']'
```

The complete source code can be found at `${HOME}/Scalable-Data-Analytics-with-Azure-Data-Explorer/Chapter04/createEplTable.kql`.

Now that we have reviewed the code, let's create the table and mapping schemas and import some data:

1. Sign in to the Data Explorer Web UI (`https://dataexplorer.azure.com/`).

2. Click **File | Open** and open `createEplTable.kql`. For instance, my file is located at `${HOME}/Scalable-Data-Analytics-with-Azure-Data-Explorer/Chapter04/createEplTable.kql`.

3. As shown in *Figure 4.9*, select the `create table` query by placing your cursor in the text. The text background will turn blue when the query is selected. Click **Run** or press the *Shift+Enter* keys to execute the query.

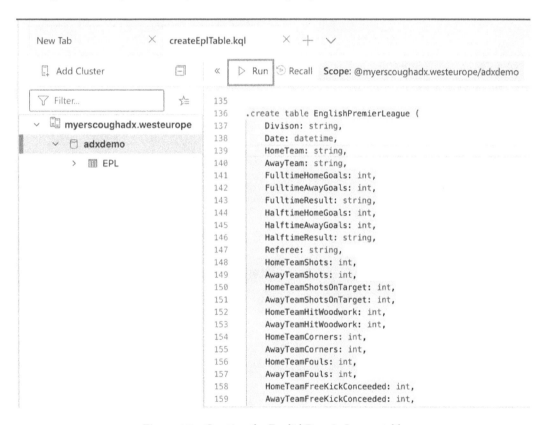

Figure 4.9 – Creating the EnglishPremierLeague table

Once the query has finished executing, the newly created table under your database is shown in the cluster pane and the table details are visible in the output pane, as shown in *Figure 4.10*:

Figure 4.10 – EnglishPremierLeague table creation summary

4.   The next step is to create our mapping schema. In the Data Explorer Web UI, select the schema mapping query, which starts on *line 170*, and click **Run**, as shown in *Figure 4.11*:

Figure 4.11 – Creating a mapping schema

Once the query has finished executing, a summary of the execution will be displayed in the output pane, as shown in *Figure 4.12*:

Figure 4.12 – Creating a mapping schema

You may notice that all the data types are set to `" "`. This is because we did not specify the data types in our mapping since the data types for our table columns were already declared in our initial `.create table` command. Now that we have created our table and mapping, let's import some data and verify whether we did indeed ignore the betting columns.

5. Ensure the scope is set to `@<your_adx_cluster>.<your_region>/<your_adxdb>`, such as `@myerscoughadx.westeurope/adxdemo`, and click **Data** from the left-hand menu. As mentioned earlier, the ingestion will fail if you are not at the database scope.

6. From **Ingest new data**, click **Ingest**.

7. Select the table we just created, `EnglishPremierLeague`, and click **Next: Source**.

8. Set **Source type** to **From file** and upload `${HOME}/Scalable-Data-Analytics-with-Azure-Data-Explorer/Chapter04/datasets/premierleague/csv/season-1718_csv.csv`.

9. Click **Next: Schema**.

    As shown in *Figure 4.13*, the data in the **Partial data preview** pane may look
    incorrect. This is because we need to use our custom mapping.

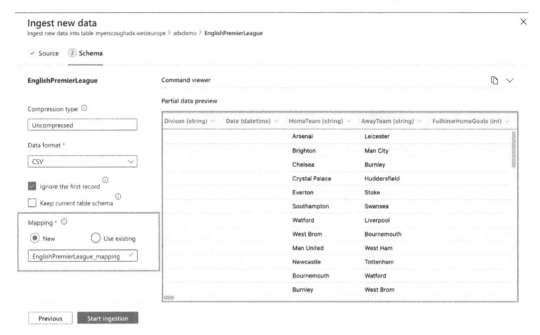

Figure 4.13 – Incorrect column mapping

10. Under **Mapping**, select **Use existing** and select our custom mapping, **EPL_Custom_
    Mapping**. Now you should see the columns mapped correctly in the Partial data
    preview, as shown in *Figure 4.14*:

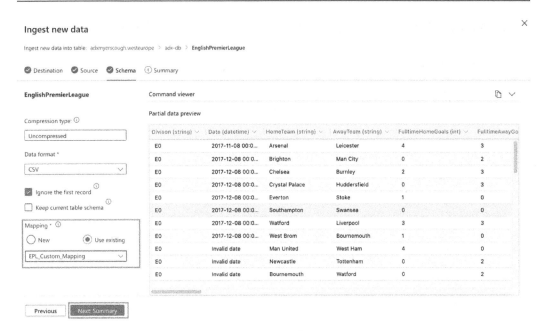

Figure 4.14 – Correct column mapping

11. Click **Next: Summary** to complete the ingestion process.

12. Once the ingestion process is complete, go back to the query editor and run the following KQL query to view all of Manchester United's results from the 2017-2018 season:

```
EnglishPremierLeague
| where HomeTeam contains "Man United" or AwayTeam
contains "Man United"
```

As shown in *Figure 4.15*, all of Manchester United's results can be seen for the 2017-2018 season, and if you scroll to the right, you will notice there are no columns related to the betting statistics:

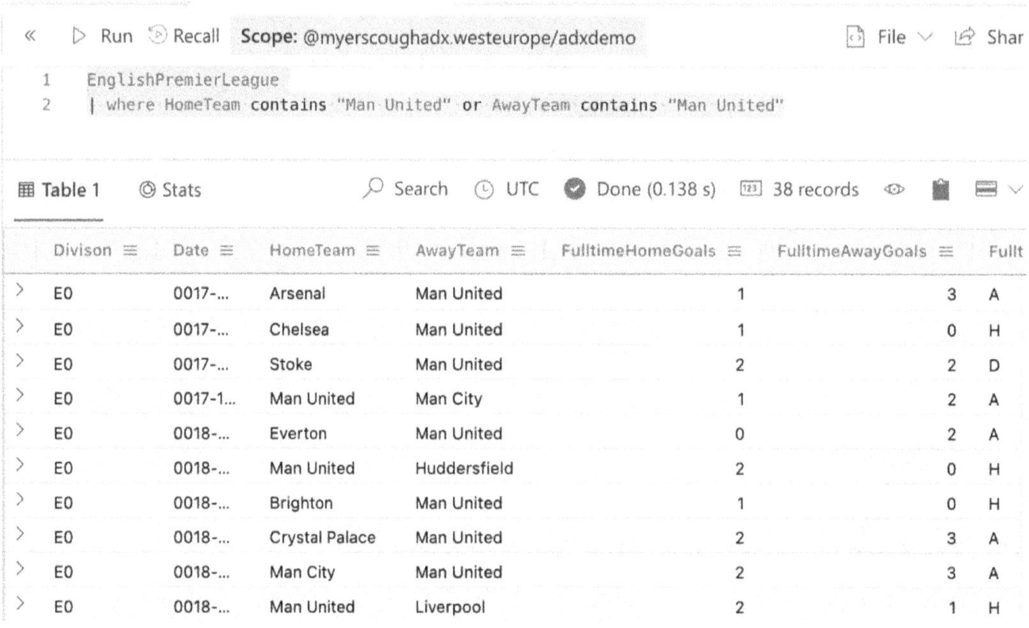
Run
Recall
Scope: @myerscoughadx.westeurope/adxdemo
File
Shar
1 EnglishPremierLeague
2 | where HomeTeam contains "Man United" or AwayTeam contains "Man United"
Table 1
Stats
Search
UTC
Done (0.138 s)
38 records
Divison
Date
HomeTeam
AwayTeam
FulltimeHomeGoals
FulltimeAwayGoals
Fullt
E0 0017-... Arsenal Man United 1 3 A
E0 0017-... Chelsea Man United 1 0 H
E0 0017-... Stoke Man United 2 2 D
E0 0017-1... Man United Man City 1 2 A
E0 0018-... Everton Man United 0 2 A
E0 0018-... Man United Huddersfield 2 0 H
E0 0018-... Brighton Man United 1 0 H
E0 0018-... Crystal Palace Man United 2 3 A
E0 0018-... Man City Man United 2 3 A
E0 0018-... Man United Liverpool 2 1 H

Figure 4.15 – Manchester United results

In this section, we created our own custom table and mapping schema to ignore some of the columns from our data source. We then imported our data and executed a query to verify the data was ingested correctly and that our table did not contain any of the betting statistics. In the next section, *Ingesting data from Blob storage using Azure Event Grid*, we will learn how to stream data using Azure Event Grid and Event Hubs.

Ingesting data from Blob storage using Azure Event Grid

In our final example of data ingestion, we will enable streaming on our cluster and use **Azure Event Grid** and **Event Hubs** so we can ingest data whenever new files are placed in our storage account's blob container. A blob container is a location on the storage account used to store our data.

For this section, we need to create the following Azure resources:

- A **storage account** for storing files
- An **event grid** to emit blob creation events
- An **event hub** deliver the notification to **Azure Data Explorer**

Using JSON data, we will demonstrate how to create JSON-based mapping schemas.

When a file is uploaded to the storage account, a *blob created* event is generated and received by the *event grid*. The *event grid* then updates *Azure Data Explorer* to pull information from the storage account. In our example, the information is a *JSON* file.

This path of data ingestion is shown in the following figure:

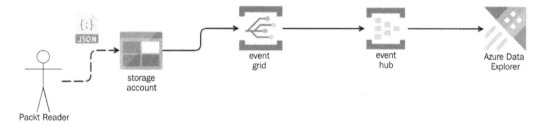

Figure 4.16 – Data ingestion using Event Grid and Event Hubs

In the following sections, we will build our infrastructure to match *Figure 4.16* and we will begin by enabling streaming on our cluster.

Enabling streaming on ADX

When streaming is enabled, data is not only sent to the column stores, but also to the row stores.

The following steps explain how to enable streaming on our ADX cluster:

1. Go to the Azure portal (`https://portal.azure.com`) and go to your ADX instance.
2. Under **Settings**, click **Configurations**, as shown in *Figure 4.17*.

3. Turn **Streaming ingestion On** and click **Save**.

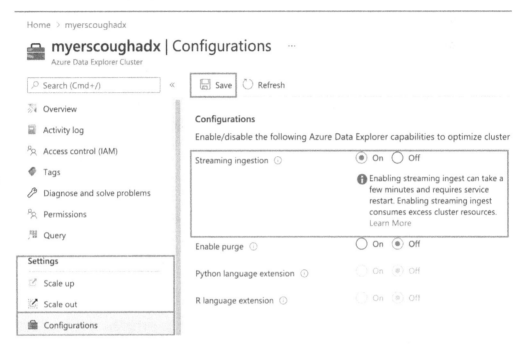

Figure 4.17 – Enabling streaming ingestion

As shown in the preceding screenshot, enabling streaming can take a couple of minutes to complete. The information prompt also mentions that streaming requires extra cluster resources, since streaming writes data to the column and row stores, whereas batching just writes to the column stores.

Now that we have enabled streaming on our cluster, let's go ahead and create our new table and mapping schema.

Creating our table and JSON mapping schema

We will now create a new table called `EnglishPremierLeagueJSON` and a mapping schema to map the date, home team, away team, and the fulltime result. *Figure 4.18* illustrates the mapping of rows in our JSON to columns in our ADX table:

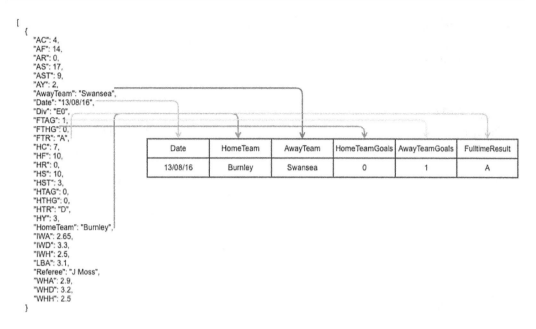

```
[
  {
    "AC": 4,
    "AF": 14,
    "AR": 0,
    "AS": 17,
    "AST": 9,
    "AY": 2,
    "AwayTeam": "Swansea",
    "Date": "13/08/16",
    "Div": "E0",
    "FTAG": 1,
    "FTHG": 0,
    "FTR": "A",
    "HC": 7,
    "HF": 10,
    "HR": 0,
    "HS": 10,
    "HST": 3,
    "HTAG": 0,
    "HTHG": 0,
    "HTR": "D",
    "HY": 3,
    "HomeTeam": "Burnley",
    "IWA": 2.65,
    "IWD": 3.3,
    "IWH": 2.5,
    "LBA": 3.1,
    "Referee": "J Moss",
    "WHA": 2.9,
    "WHD": 3.2,
    "WHH": 2.5
  }
}
```

Date	HomeTeam	AwayTeam	HomeTeamGoals	AwayTeamGoals	FulltimeResult
13/08/16	Burnley	Swansea	0	1	A

Figure 4.18 – JSON mapping schema

Let's create our table using the following KQL management commands:

```
.create table EnglishPremierLeagueJSON (
    Date: string,
    HomeTeam: string,
    AwayTeam: string,
    FulltimeHomeGoals: int,
    FulltimeAwayGoals: int,
    FulltimeResult: string
)
```

You may have noticed that the Date column is a string rather than being of the Date type. This is because the data is not in a date format that is recognized by ADX, which causes problems when trying to load the values as Date.

The following sequence of steps will demonstrate how to create a new table called
EnglishPremierLeagueJSON, which we will use to store the JSON-based data
we will be ingesting:

1. Log in to https://dataexplorer.azure.com.

2. Click **Add Cluster** and enter the name of your cluster, for example,
   https://adxmyerscough.westeurope.kusto.windows.net.

3. Select the database scope by expanding the cluster and clicking on your database.

4. Click **File** and **Open** and select ${HOME}/Scalable-Data-Analytics-
   with-Azure-Data-Explorer/Chapter04/createEplJSONTable.kql,
   as shown in *Figure 4.19*:

```
« ▷ Run ⟳ Recall  Scope: @adxmyerscough.westeurope/adx-db

1
2
3    // Creates a new table called EnglishPremierLeagueJSON
4    .create table EnglishPremierLeagueJSON (
5        Date: string,
6        HomeTeam: string,
7        AwayTeam: string,
8        FulltimeHomeGoals: int,
9        FulltimeAwayGoals: int,
10       FulltimeResult: string
11   )
12
13   // Creates a new JSON mapping that maps the JSON values to our table's columns
14   .create table ['EnglishPremierLeagueJSON'] ingestion json mapping 'ELP_Custom_JSON_Mapping'
15   '['
16       '{"column": "Date", "Properties": {"Path": "$.Date"}},'
17       '{"column": "HomeTeam", "Properties": {"Path": "$.HomeTeam"}},'
18       '{"column": "AwayTeam", "Properties": {"Path": "$.AwayTeam"}},'
19       '{"column": "FulltimeHomeGoals", "Properties": {"Path": "$.FTHG"}},'
20       '{"column": "FulltimeAwayGoals", "Properties": {"Path": "$.FTAG"}},'
21       '{"column": "FulltimeResult", "Properties": {"Path": "$.FTR"}},'
22   ']'
23
```

Figure 4.19 – Creating our new table and mapping schema

5. Move your cursor to the first query and click **Run**. Once the query has executed,
   you will see the EnglishPremierLeagueJSON table below your database in
   the cluster pane.

6. Next, move your cursor to the second query and click **Run**. The second query will
   create our JSON mapping schema called EPL_Custom_JSON_Mapping.

Now that we have created our table and mapping schema in ADX, let's start adding the additional infrastructure by starting with the storage account where we will upload the data we want to be ingested into ADX.

Creating our storage account

We will deploy our storage account using our **Azure Resource Manager** (**ARM**) template, which can be found at `${HOME}/Scalable-Data-Analytics-with-Azure-Data-Explorer/Chapter04/templates/storageaccount.json`.

The following steps will explain how to deploy a new storage account using our ARM templates:

1. Open **Azure Cloud Shell** by going to `https://shell.azure.com`.
2. Navigate to the templates directory by typing `cd ${HOME}/Scalable-Data-Analytics-with-Azure-Data-Explorer/Chapter04/templates`, where `${HOME}` is the parent directory of `Scalable-Data-Analytics-with-Azure-Data-Explorer/`.
3. Storage account names must be globally unique. So let's edit the storage account parameter file and select a name that will be globally unique. Type `code .` to open the lightweight Visual Studio Code editor.
4. Update *line 6* and choose a name that is unique. There is a good chance another reader has deployed a storage account called `packtdemo`. Click the *Ctrl + S* keys on Windows or the *Cmd + S* keys if you are on macOS to save your changes.
5. Deploy the new storage account by typing `New-AzResourceGroupDeployment -Name "StorageDeployment" -ResourceGroup "adx-rg" -TemplateFile ./storageaccount.json -TemplateParameterFile ./storageaccount.params.json`.

    As you can see, we are deploying the storage account into the resource group where we have deployed our ADX cluster.

6. Once the deployment is complete, a summary of the deployment is displayed in the console, as shown in *Figure 4.20*:

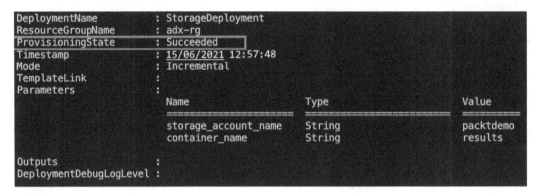

Figure 4.20 – Storage account deployment summary

When we are ready to start ingestion data, we will upload our JSON files to our storage account. In the next section, *Creating our event hub*, we will deploy our ARM templates for the event hub.

Creating our event hub

In this section, we will deploy an Azure Event Hubs namespace and event hub using our ARM templates.

The following steps will explain how to deploy an event hub using our ARM templates:

1. Please follow *steps 1 and 2* from the *Creating our storage account* section.

2. Since event hub names need to be globally unique, let's first update the parameter file and change the name of the event hub we plan to deploy. Type code . to open the lightweight Visual Studio Code editor.

3. As shown in *Figure 4.21*, update *lines 6 and 9* and name your event hub as something that is globally unique:

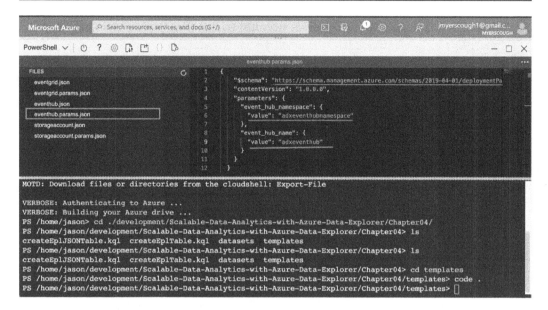

Figure 4.21 – Updating the event hub name

4. If you are using Windows, press *Ctrl + S* or if you are on a Mac, press *Cmd + S* to save your changes.

5. Deploy the new event hub by typing `New-AzResourceGroupDeployment -Name "EventHubDeployment" -ResourceGroup "adx-rg" -TemplateFile ./eventhub.json -TemplateParameterFile ./eventhub.params.json`. As you can see, we are deploying the event hub into the resource group where we have deployed our ADX cluster.

With our event hub deployed, the last step before we start ingesting data is to create an Event Grid, which we will create in the next section, *Creating our Event Grid*.

Creating our Event Grid

The following steps will explain how to deploy and configure an Event Grid using the ARM templates:

1. Please follow *steps 1 and 2* from the *Creating our storage account* section.

2. If you changed the name of the event hub in the previous section, then you need to update *lines 6 and 9* in `eventgrid.params.json` to match.

3. If you changed the name of the storage account earlier, then you will also need to update *line 15* in `eventgrid.params.json` to match.

4. Deploy the new Event Grid by typing `New-AzResourceGroupDeployment -Name "EventGridDeployment" -ResourceGroup "adx-rg" -TemplateFile ./eventgrid.json -TemplateParameterFile ./ eventgrid.params.json`. As you can see, we are deploying the Event Grid into the resource group where we have deployed our ADX cluster.

With our infrastructure deployed, we can now prepare to ingest data from our blob container. In the next section, *Ingesting data in ADX*, we will create a data connection between the storage account and the ADX cluster and ingest our JSON data by uploading the files to the blob container.

Ingesting data in ADX

In this section, we are going to create our data connection and upload our JSON files to the blob container and when the *blob created* event is triggered, our files will be ingested into ADX.

Before proceeding, you should be aware that ADX supports two types of **JSON**: **JSON** and **Multiline JSON**, and it is important you understand the difference. If you select the incorrect type, then your data will not be ingested.

The first method, called JSON, expects the complete record to be on one line, for example:

```
{"name": "harrison", "dob": "17/04/2012"}

{"name": "james", "dob": "09/07/2014"}
```

Multiline JSON allows records to span multiple lines and the parser will ignore the whitespace, for example:

```
{
    "name": "harrison",
    "dob": "17/04/2012"
},
{
    "name": "james",
    "dob": "09/07/2014"
}
```

The following steps explain how to create a data connection and how to upload files to the blob storage:

1. In the Azure portal, click on **Azure Data Explorer** and, from the **Properties** pane, select **Databases**, as shown in *Figure 4.22*:

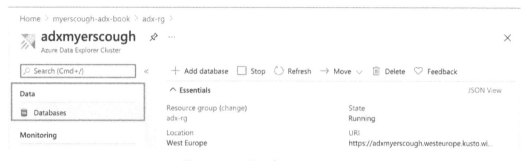

Figure 4.22 – Database properties

2. Select the **adx-db** database.

3. Click **Data connections** from the **Settings** pane and then click **+Add data connection**, as shown in *Figure 4.23*:

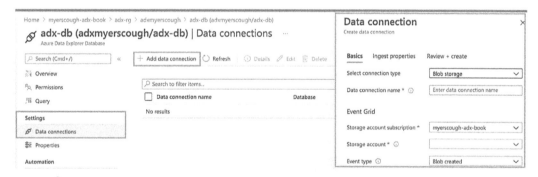

Figure 4.23 – Creating a data connection

4. For **Select connection type**, select **Blob storage**.

5. For **Data connection name**, use `adxBlobConnection`.

6. Select the storage account – select the **packtdemo** storage account or the name of your storage account if you changed it earlier.

7. Under **Resource creation**, select **Manual**. **Automatic** will take care of creating the Event Grid and event hub for us.

8. Select **adxEventGridSubscription** as the **Event Grid**. The event hub name will automatically be set to adxeventhub or the name you set in the parameter file earlier, which we created in the previous section, *Creating our event hub,* with our ARM templates.

9. Select **$Default** for **Consumer group**, as shown in *Figure 4.24,* and then click **Next: Ingest properties >**:

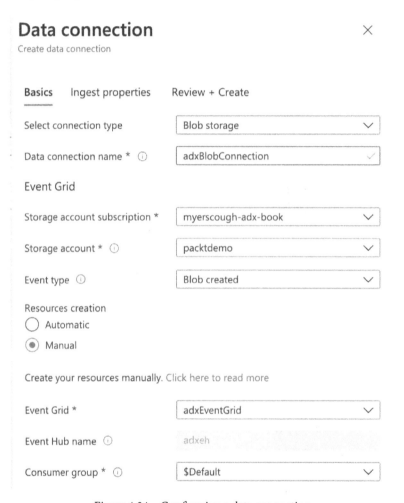

Figure 4.24 – Configuring a data connection

10. Next, enter the name of our table: `EnglishPremierLeagueJSON`.

11. Select the **Data format** option as **MULTILINE JSON**.

12. Select the **Mapping name** option for our JSON mapping schema as **EPL_Custom_ JSON_Mapping**, and then click **Next: Review + create**.

13. Once the validation has completed, click **Create**.

Once the data connection has been created, we can begin to ingest data by uploading the data to our blob container. The following sequence of steps explains how to ingest JSON data by uploading the files to our blob storage:

1. In the Azure portal, open the `packtdemo` storage account or whichever name you gave it earlier if you changed the parameter file.

2. Under the **Data Storage** properties, select **Containers** and then click our **Results** container.

3. Select **Upload** and upload one of our JSON files, such as `${HOME}/Scalable- Data-Analytics-with-Azure-Data-Explorer/Chapter04/ datasets/premierleague/json/season-1516_json.json`. After a couple of minutes, the data will be available in our table, ready to be queried, as shown in *Figure 4.26*. The duration you must wait before seeing the data depends on the configuration of the batching policy. By default, the policy is set to 5 minutes.

   As shown in *Figure 4.25*, the event hub metrics show one incoming and one outgoing message. These messages are the result of our file upload to the blob container.

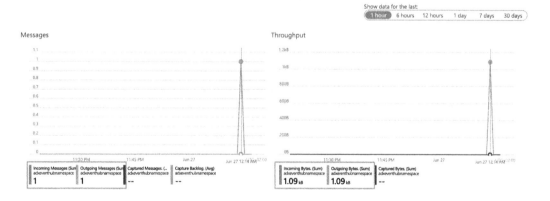

Figure 4.25 – Event hub metrics

Once the data has been ingested, you can query the table as shown in *Figure 4.26*:

```
23
24    EnglishPremierLeagueJSON
25    | limit 50
```

⊞ Table 1   ◎ Stats

Date ≡	HomeTeam ≡	AwayTeam ≡	FulltimeHomeGoals ≡	FulltimeAwayGoals ≡	FulltimeResult ≡
10/08/...	Man United	Leicester	2	1	H
11/08/...	Bournemouth	Cardiff	2	0	H
11/08/...	Fulham	Crystal Palace	0	2	A
11/08/...	Huddersfield	Chelsea	0	3	A
11/08/...	Newcastle	Tottenham	1	2	A
11/08/...	Watford	Brighton	2	0	H
11/08/...	Wolves	Everton	2	2	D
12/08/...	Arsenal	Man City	0	2	A
12/08/...	Liverpool	West Ham	4	0	H
12/08/...	Southampton	Burnley	0	0	D
18/08/...	Cardiff	Newcastle	0	0	D
18/08/...	Chelsea	Arsenal	3	2	H

Figure 4.26 – Querying EnglishPremierLeagueJSON

Every time a file is created on the blob container, a *blob created* event will be triggered, which will result in the file being ingested into our table. You can try to upload another file and see whether the data has been ingested into the table.

Summary

This chapter covered a lot of topics to ingest… I mean digest – pardon the pun. We started by learning about data ingestion in general, discussing the different types of data, such as structured, semi-structured, and unstructured. We then looked at the data management service in more detail to understand its role with regard to data ingestion. We also looked at the difference between batching and streaming data. We introduced the main ingestion categories: SDKs, managed pipelines such as Azure Event Grid, connections and plugins, and tools such as Azure Data Factory and one-click ingestion.

Then we learned about schema mapping, how they map external data to the columns in our ADX tables, and how to write our own schema maps for both CSV and JSON data. We created two schema maps for the English Premier League football results data, one for the CSV-based data and one for the JSON data, where we excluded the betting information and kept the actual football match statistics.

Finally, we learned how to ingest data using a variety of methods. We first looked at one-click ingestion where we simply uploaded our data via the Data Explorer Web UI and let ADX determine the schema mapping for us. The one-click ingestion approach is a handy method for reverse engineering datasets to determine the schema, which can be refined later if necessary. We also learned how to ingest data using KQL management commands and how to apply our custom schema mappings, and finally, we learned how to ingest data from blob containers using Azure Event Hubs and Event Grid.

In the next chapter, *Introducing KQL*, we will learn about Kusto Query Language so we can start querying our data, and begin to look for trends and anomalies in our datasets.

Questions

Before moving on to the next chapter, test your knowledge by trying these exercises. The answers can be found at the back of the book.

1. Which is the preferred ingestion method – streaming or batch ingestion?
2. Try to import the StormEvents CSV file using the one-click ingestion method. We will use the StormEvents table in the next chapter.
3. Try to upload another Premier league JSON file.
4. Update EPL_Custom_JSON_Mapping and include the referee.

5

Introducing the Kusto Query Language

By this point, you have a solid understanding of what **Azure Data Explorer** (**ADX**) is, how to use it, and how to build ADX infrastructure via the **Azure portal**. You also know how to use **Microsoft PowerShell** and **Azure Resource Manager** (**ARM**) templates, and how to configure data ingestion. Now that we have ingested data, the next step is to understand how to query and explore our data. In this chapter, we are going to introduce the **Kusto Query Language** (**KQL**), and then in the next two chapters, *Chapter 6, Introducing Time Series Analysis*, and *Chapter 7, Identifying Patterns, Anomalies, and Trends in Your Data*, we will focus on the advanced features of KQL.

We will begin by explaining what KQL is, what its main features are, and where KQL can be used. Next, we will learn about the syntax and structure of KQL queries, as well as how to search using the `search` and `where` operators. Then, we will explore how to perform aggregation using the `summarize` operator, and introduce the `bin()` function to bucket our values. We will also learn how to convert **SQL** statements to KQL queries using the `EXPLAIN` keyword.

Next, we will learn about the arithmetic, logical, and relational operators that will help us write complex search predicates, and we will introduce some of the string search operators and learn how to manipulate date and time values. After that, we will learn about joining tables and the different types of joins that are available.

Finally, we will introduce some of KQL's management commands. Like SQL, KQL has two categories of commands – *data manipulation commands*, which allow us to query our data, and *management commands*, which allow us to manage and connect to our clusters and databases.

In this chapter, we are going to cover the following main topics:

- What is KQL?
- Introducing the basics of KQL
- Introducing KQL's scalar operators
- Joining tables in KQL
- Introducing KQL's management commands

Technical requirements

The code examples for this chapter can be found in the Chapter05 folder of our repository:

https://github.com/PacktPublishing/Scalable-Data-Analytics-with-Azure-Data-Explorer.git

> **Note**
>
> Since KQL is used in a variety of places in Azure, we will use https://aka.ms/LADemo in some of our examples. The log analytics demo provides us with a sample set of common tables in Azure, such as Performance, Security Center, Azure Sentinel, and so on. This provides you with some exposure to the tables that engineers use on a regular basis when using Azure.

What is KQL?

Whenever someone asks me about the most important features of Azure to learn, one of the first features I mention is **KQL**. KQL is the language for managing all your logging and telemetry on Azure. Even if you do not run and manage an ADX cluster, you will still need KQL for monitoring, analyzing your logs, managing your assets, exploring your security data, and exploring your **ADX Application Insights** data.

KQL is ADX's read-only query language that has many similarities with SQL, such as working with tables, columns, and providing functionality for filtering. For those who are not familiar with SQL, it is a query language for relational databases. As we will see later in this book, KQL supports a subset of SQL, and SQL statements can be executed and converted to KQL using the EXPLAIN keyword, reducing the learning curve for engineers with an SQL background.

KQL is not only used directly in ADX and the ADX Web **user interface** (UI). If you recall *Chapter 1*, *Introducing Azure Data Explorer*, one of Azure's services, **Log Analytic Workspace**, is built on top of ADX, and Azure uses it for storing all logging and telemetry data, such as that used in **Monitoring**, **Security Center**, **Update Management**, **Inventory Management**, and **Application Insights**. This data can be queried using KQL.

Therefore, if you find yourself not managing ADX clusters and databases daily, there is still a high probability you will still use KQL to manage all your logging and telemetry data.

Next, let's look at the syntax for a KQL query. *Figure 5.1* depicts the syntax for KQL queries:

```
// comments
Table  |  Data Transformation Statement 1  |  ...  |  Data Transformation Statement n  |  render statement
```

Figure 5.1 – KQL query syntax

The // comments are optional, but it is good practice to write comments to explain what the query is doing. A lot of the complex query examples in this book have a comment section that explains what the query does.

The next section is Table, which is the data source that we want to query. Strictly speaking, the data source does not need to be a table – it can be an expression or a nested KQL query.

The next element is a pipe (|) followed by one or more data transformation statements. For those familiar with working in a command prompt such as PowerShell or **Bash**, pipes are used to pass standard-input and standard-output to commands. The concept is similar in KQL.

We can read the example in *Figure 5.1* as follows: take Table and pipe it to Data Transformation Statement 1, which performs some operation (such as filtering) that returns a dataset as the result. This result set is then piped to the subsequent data transformation statements via the pipe (|).

The final element of a query, render statement, which is optional, is used to generate graphs. We will learn how to render graphs in the next section, *Introducing the basics of KQL*.

When you write queries in the ADX Web UI or in the Log Analytics query editor, you place your cursor on the query and then click **Run**, or press *Shift + Enter*. This means that the queries are executed one at a time. It is possible to execute multiple queries during one execution by chaining them with a semi-colon. For example, the following query creates a variable called `stateToSearch` and returns the `State` and `EventType` columns.

As you can see in the `where` clause, the variable is used in the search expression:

```
let stateToSearch = 'FLORIDA';
StormEvents
| where State == stateToSearch
| project State, EventType
```

We will learn more about variables in the *Introducing the basics of KQL* section.

As a concrete example, imagine we have a table called `Family` that has two columns, `Name` and `Gender`, as shown in *Table 5.1*:

Name	Gender
Harrison	Male
James	Male
Lukas	Male
Damian	Male
Diana	Female
Sofia	Female
Charlotte	Female
Dave	Male

Table 5.1 – Family table

The following query returns a list of all the males that have a D in their name.

```
Family
| where Gender == "Male"
| where Name contains "D"
```

The first line, `Family`, pipes the entire dataset to the first `where` statement (`where Gender == "Male"`). The result of this `where` clause is all the males, as shown in *Table 5.2*:

Name	Gender
Harrison	Male
James	Male
Lukas	Male
Damian	Male
Dave	Male

Table 5.2 – Result containing all males

This intermediate result set is then piped to the second `where` clause (`where Name contains "D"`), which then returns the result, as shown in *Table 5.3*:

Name	Gender
Damian	Male
Dave	Male

Table 5.3 – Final result of the Family query

> **Note**
>
> If you want to execute the preceding query, I have included the dataset and query in the GitHub repository. You can find the files in the `Chapter05` directory: `${HOME}/Scalable-Data-Analytics-with-Azure-Data-Explorer/Chapter05/family.kql` and `${HOME}/Scalable-Data-Analytics-with-Azure-Data-Explorer/Chapter05/datasets/family.csv`
>
> The dataset can be imported into your cluster using the *one-click ingestion* method that we discussed in *Chapter 4, Ingesting Data in Azure Data Explorer*.

In the next section, we will learn how to search for data, perform basic aggregations, and render graphs.

Introducing the basics of KQL

Throughout this chapter, I will draw comparisons between SQL and KQL to demonstrate similarities and showcase the simplicity of KQL. Before we start to look at the basic data transformation operators, let's first look at how to query a table in the simplest form.

In SQL, if you want to query a table and return all columns and rows, you can execute a query as follows:

```
Select * from StormEvents
```

The query returns all the rows and columns for `StormEvents`. You can even execute the SQL query in the ADX Web UI. The equivalent query in KQL is simply the table name:

```
StormEvents
```

As shown in *Figure 5.2*, the query returns all rows and columns (`59,066` records) in approximately 23 seconds:

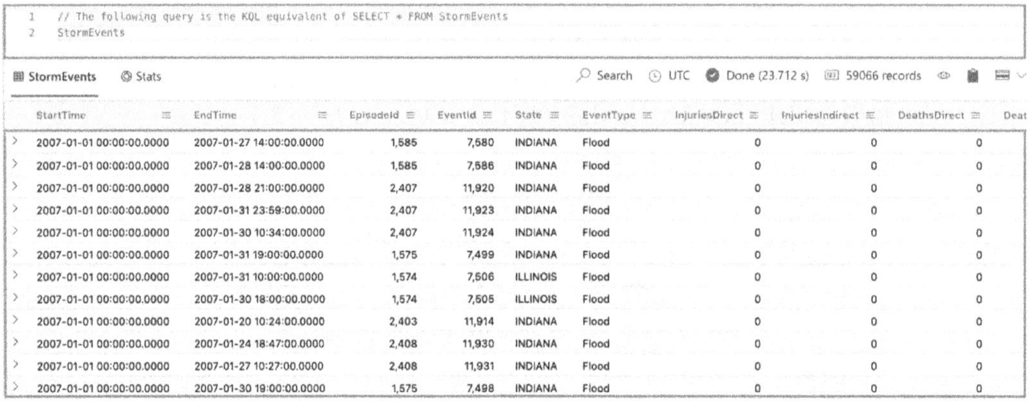

StartTime	EndTime	EpisodeId	EventId	State	EventType	InjuriesDirect	InjuriesIndirect	DeathsDirect	Deat
2007-01-01 00:00:00.0000	2007-01-27 14:00:00.0000	1,585	7,580	INDIANA	Flood	0	0	0	
2007-01-01 00:00:00.0000	2007-01-28 14:00:00.0000	1,585	7,586	INDIANA	Flood	0	0	0	
2007-01-01 00:00:00.0000	2007-01-28 21:00:00.0000	2,407	11,920	INDIANA	Flood	0	0	0	
2007-01-01 00:00:00.0000	2007-01-31 23:59:00.0000	2,407	11,923	INDIANA	Flood	0	0	0	
2007-01-01 00:00:00.0000	2007-01-30 10:34:00.0000	2,407	11,924	INDIANA	Flood	0	0	0	
2007-01-01 00:00:00.0000	2007-01-31 19:00:00.0000	1,575	7,499	INDIANA	Flood	0	0	0	
2007-01-01 00:00:00.0000	2007-01-31 10:00:00.0000	1,574	7,506	ILLINOIS	Flood	0	0	0	
2007-01-01 00:00:00.0000	2007-01-30 18:00:00.0000	1,574	7,505	ILLINOIS	Flood	0	0	0	
2007-01-01 00:00:00.0000	2007-01-20 10:24:00.0000	2,403	11,914	INDIANA	Flood	0	0	0	
2007-01-01 00:00:00.0000	2007-01-24 18:47:00.0000	2,408	11,930	INDIANA	Flood	0	0	0	
2007-01-01 00:00:00.0000	2007-01-27 10:27:00.0000	2,408	11,931	INDIANA	Flood	0	0	0	
2007-01-01 00:00:00.0000	2007-01-30 19:00:00.0000	1,575	7,498	INDIANA	Flood	0	0	0	

Figure 5.2 – The simplest KQL query

This type of query can be expensive in terms of performance since it returns all records and columns, and tables can contain millions of records. This type of query is normally used with `limit`. The `limit` operator is used to limit the number of returned rows. For example, the following query will return up to 10 records:

```
StormEvents
| limit 10
```

This type of query is convenient as a quick means of validating whether your table contains data.

> **Note**
>
> For now, do not worry too much about performance and let's focus on learning the basics of KQL. To learn more about query performance, see *Chapter 11, Performance Tuning in Azure Data Explorer*.

In practice, you typically want to search for specific values or conditions when querying your data.

In the next section, we will learn what search predicates are and how they are used in queries.

Introducing predicates

Before jumping into KQL, it is important to understand what search predicates are, as you will see this term a lot throughout this chapter and in Microsoft's documentation. *Predicates*, with regard to KQL and programming in general, are expressions that evaluate to a Boolean value – that is, either `true` or `false`. For example, as we will learn in the next section, *Searching and filtering*, the `search` operator can search across all tables and columns. The syntax for the `search` operator, when used with a table, is `Table | search Predicate`. If the predicate is `true`, the record will be added to the result set.

For example, consider the following query:

```
StormEvents
| search "thunder"
```

The query searches all columns in the `StormEvents` table. If the query finds the `thunder` string, the expression or predicate evaluates to `true` and the row is added to the result set.

Now that we understand the basics of KQL syntax and predicates, in the next section, we will learn how to search our data.

Searching and filtering data

We will begin by looking at the `search` operator, which is not the most efficient when it comes to searching for values, but is convenient when working on small datasets. The preferred method is the `where` operator, which we will also learn about in this section.

The search operator

KQL provides the `search` operator to search for text literals, regular expressions, and Boolean expressions. The `search` operator works by searching across multiple tables and all columns, which, as you can imagine, is intensive on compute resources when dealing with large datasets and is not recommended. You can restrict the scope of the `search` operator with some additional arguments, as we will see shortly.

The syntax for the search operator is as follows:

```
[Table |] search [kind=CaseSensitivity] [in (TableSource)
SearchPredicate
```

The square brackets ([]) denote the optional elements of the `search` operator. As you can see, `Table` is optional. If you do not specify a table, the `search` operator will search across all tables in the database.

The following query is valid and will search across all the tables and all columns within your database:

```
search "Thunderstorm"
```

By default, `search` is case-insensitive, meaning, the preceding query would return `Thunderstorm, thunderstorm`.

As shown in *Figure 5.3*, the query returned records containing the word `Thunderstorm` and `thunderstorms`:

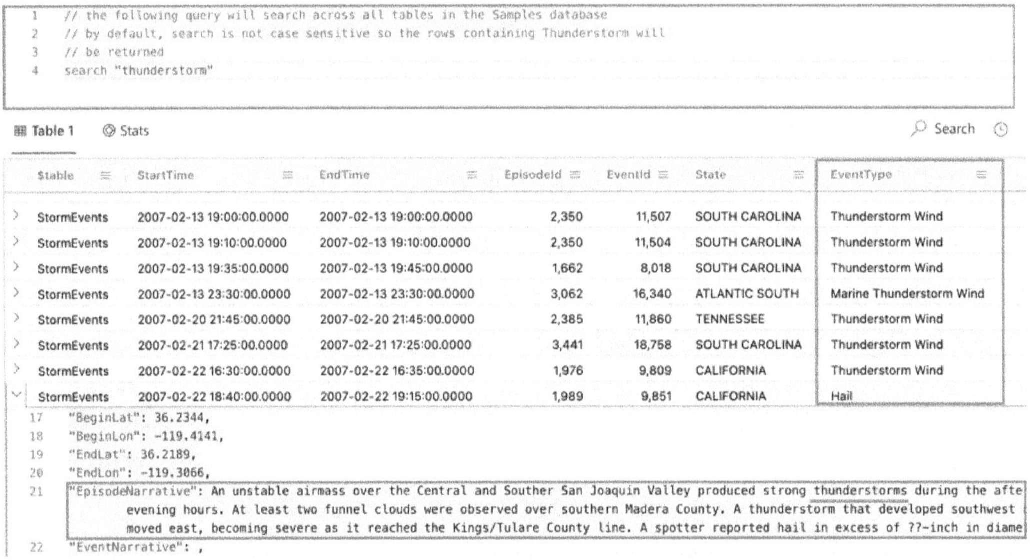

Figure 5.3 – Searching across all tables and columns

In the `EventType` column, you can see that there are a couple of records that do not contain the word `Thunderstorm` (such as `Hail`) but are still displayed as part of the result set. This is because the `search` operator is searching across all columns and in *Figure 5.3*, the `EpisodeNarrative` column contains the word `thunderstorms`:

In practice, you typically want to search for specific values or conditions when querying your data.

In the next section, we will learn what search predicates are and how they are used in queries.

Introducing predicates

Before jumping into KQL, it is important to understand what search predicates are, as you will see this term a lot throughout this chapter and in Microsoft's documentation. *Predicates*, with regard to KQL and programming in general, are expressions that evaluate to a Boolean value – that is, either `true` or `false`. For example, as we will learn in the next section, *Searching and filtering*, the `search` operator can search across all tables and columns. The syntax for the `search` operator, when used with a table, is `Table | search Predicate`. If the predicate is `true`, the record will be added to the result set.

For example, consider the following query:

```
StormEvents
| search "thunder"
```

The query searches all columns in the `StormEvents` table. If the query finds the `thunder` string, the expression or predicate evaluates to `true` and the row is added to the result set.

Now that we understand the basics of KQL syntax and predicates, in the next section, we will learn how to search our data.

Searching and filtering data

We will begin by looking at the `search` operator, which is not the most efficient when it comes to searching for values, but is convenient when working on small datasets. The preferred method is the `where` operator, which we will also learn about in this section.

The search operator

KQL provides the `search` operator to search for text literals, regular expressions, and Boolean expressions. The `search` operator works by searching across multiple tables and all columns, which, as you can imagine, is intensive on compute resources when dealing with large datasets and is not recommended. You can restrict the scope of the `search` operator with some additional arguments, as we will see shortly.

The `where` operator expects a data source (for example, a table) to be piped to it. The following query searches the `StormEvents` table for `DamageProperty > 0`:

```
StormEvents
| where DamageProperty > 0
```

KQL also supports compound expressions using logical operators. We will cover *arithmetic*, *logical* and *relational* operators in *Introducing KQL's scalar operators*. The following query checks for `DamageProperty > 0` in Indiana:

```
StormEvents
| where State == 'INDIANA' and DamageProperty > 0
```

As shown in *Figure 5.4*, the query returns records where the `State` is equal to `INDIANA` and the property damage is greater than `0`:

```
12    // This query searches for Property damage in Indiana that is
13    // greater than 0
14    StormEvents
15    | where State == 'INDIANA' and DamageProperty > 0
```

| ▦ Table 1 | ⊚ Stats | | | 🔍 Search | 🕐 UTC | ✅ Done (0.483 s) | 🔢 376 records | 👁 | 📋 | 🖥 ⌄ |

Id ≡	State ≡	EventType ≡	InjuriesDirect ≡	InjuriesIndirect ≡	DeathsDirect ≡	DeathsIndirect ≡	DamageProperty ≡	DamageCrigeCr⸳
11,920	INDIANA	Flood	0	0	0	0	10,000	
11,923	INDIANA	Flood	0	0	0	0	10,000	
11,924	INDIANA	Flood	0	0	0	0	10,000	
11,914	INDIANA	Flood	0	0	0	0	10,000	

Figure 5.4 – Boolean operators used to create compound statements

The `and` operator requires both the left-hand side and the right-hand side to be `true`. If either side is `false`, the overall expression evaluates to `false`.

We will learn more about the KQL operators we can use in our expressions in *Introducing KQL's scalar operators*. Next, let's continue with the basics and look at the basic operators for aggregation in the next section.

Aggregating data and tables

In this section, we will begin to learn about *data aggregation*, and how we can perform data aggregation using the `summarize` operator. Before diving into the `summarize` operator, we should understand the `count` operator, which is commonly used with `summarize`. The `count` operator simply counts the number of rows. For example, the following query returns the total number of rows in the `StormEvents` tables:

```
StormEvents
| count
```

In practice, you typically want to search for specific values or conditions when querying your data.

In the next section, we will learn what search predicates are and how they are used in queries.

Introducing predicates

Before jumping into KQL, it is important to understand what search predicates are, as you will see this term a lot throughout this chapter and in Microsoft's documentation. *Predicates*, with regard to KQL and programming in general, are expressions that evaluate to a Boolean value – that is, either `true` or `false`. For example, as we will learn in the next section, *Searching and filtering*, the `search` operator can search across all tables and columns. The syntax for the `search` operator, when used with a table, is `Table | search Predicate`. If the predicate is `true`, the record will be added to the result set.

For example, consider the following query:

```
StormEvents
| search "thunder"
```

The query searches all columns in the `StormEvents` table. If the query finds the `thunder` string, the expression or predicate evaluates to `true` and the row is added to the result set.

Now that we understand the basics of KQL syntax and predicates, in the next section, we will learn how to search our data.

Searching and filtering data

We will begin by looking at the `search` operator, which is not the most efficient when it comes to searching for values, but is convenient when working on small datasets. The preferred method is the `where` operator, which we will also learn about in this section.

The search operator

KQL provides the `search` operator to search for text literals, regular expressions, and Boolean expressions. The `search` operator works by searching across multiple tables and all columns, which, as you can imagine, is intensive on compute resources when dealing with large datasets and is not recommended. You can restrict the scope of the `search` operator with some additional arguments, as we will see shortly.

The syntax for the search operator is as follows:

```
[Table |] search [kind=CaseSensitivity] [in (TableSource)
SearchPredicate
```

The square brackets ([]) denote the optional elements of the `search` operator. As you can see, `Table` is optional. If you do not specify a table, the `search` operator will search across all tables in the database.

The following query is valid and will search across all the tables and all columns within your database:

```
search "Thunderstorm"
```

By default, `search` is case-insensitive, meaning, the preceding query would return `Thunderstorm, thunderstorm`.

As shown in *Figure 5.3*, the query returned records containing the word `Thunderstorm` and `thunderstorms`:

```
1  // the following query will search across all tables in the Samples database
2  // by default, search is not case sensitive so the rows containing Thunderstorm will
3  // be returned
4  search "thunderstorm"
```

$table	StartTime	EndTime	EpisodeId	EventId	State	EventType
StormEvents	2007-02-13 19:00:00.0000	2007-02-13 19:00:00.0000	2,350	11,507	SOUTH CAROLINA	Thunderstorm Wind
StormEvents	2007-02-13 19:10:00.0000	2007-02-13 19:10:00.0000	2,350	11,504	SOUTH CAROLINA	Thunderstorm Wind
StormEvents	2007-02-13 19:35:00.0000	2007-02-13 19:45:00.0000	1,662	8,018	SOUTH CAROLINA	Thunderstorm Wind
StormEvents	2007-02-13 23:30:00.0000	2007-02-13 23:30:00.0000	3,062	16,340	ATLANTIC SOUTH	Marine Thunderstorm Wind
StormEvents	2007-02-20 21:45:00.0000	2007-02-20 21:45:00.0000	2,385	11,860	TENNESSEE	Thunderstorm Wind
StormEvents	2007-02-21 17:25:00.0000	2007-02-21 17:25:00.0000	3,441	18,758	SOUTH CAROLINA	Thunderstorm Wind
StormEvents	2007-02-22 16:30:00.0000	2007-02-22 16:35:00.0000	1,976	9,809	CALIFORNIA	Thunderstorm Wind
StormEvents	2007-02-22 18:40:00.0000	2007-02-22 19:15:00.0000	1,989	9,851	CALIFORNIA	Hail

```
17  "BeginLat": 36.2344,
18  "BeginLon": -119.4141,
19  "EndLat": 36.2189,
20  "EndLon": -119.3066,
21  "EpisodeNarrative": An unstable airmass over the Central and Souther San Joaquin Valley produced strong thunderstorms during the afte
        evening hours. At least two funnel clouds were observed over southern Madera County. A thunderstorm that developed southwest
        moved east, becoming severe as it reached the Kings/Tulare County line. A spotter reported hail in excess of ??-inch in diame
22  "EventNarrative": ,
```

Figure 5.3 – Searching across all tables and columns

In the `EventType` column, you can see that there are a couple of records that do not contain the word `Thunderstorm` (such as `Hail`) but are still displayed as part of the result set. This is because the `search` operator is searching across all columns and in *Figure 5.3*, the `EpisodeNarrative` column contains the word `thunderstorms`:

As mentioned earlier, searching across all tables and columns is expensive. There are a few ways we can limit the tables being searched. The first method is to pipe the table to the `search` operator, as follows:

```
StormEvents
| search "thunderstorm"
```

The `search` operator still searches across all columns but is restricted to just the `StormEvents` table. The second method is to use the `in` argument. For example, the following query limits the scope of `search` to the `StormEvents` and `PopulationData` tables:

```
search in (StormEvents, PopulationData) "thunderstorm"
```

The `in` operator takes a list of tables or data sources and limits the search to those tables. The `search` operator also allows us to restrict the columns that it will search by specifying the column name followed by a colon, followed by the search term. For example, the following query searches all tables but only searches the `EventType` column:

```
search EventType:"thunderstorm wind"
```

The `search` operator does not just work with string predicates – any predicate can be used. For example, the following predicate searches the `DamageProperty` column for values greater than `0`:

```
StormEvents
| search DamageProperty > 0
```

Although the `search` operator is simple and convenient to use, you need to be careful with regard to the scope of the search. Searching across multiple tables and columns is an expensive operation and can cause performance issues. In the next section, we are going to look at another operator, called `where`, that allows us to search and filter our data, and is the preferred approach.

The where operator

The `where` operator is similar to the `WHERE` clause in SQL, whereby you write an expression to search for a condition in your data. If the expression evaluates to `true`, the row is added to the result set. Unlike the `search` operator, the `where` operator does not work globally across tables, and the syntax for using where is as follows:

```
Table | where Predicate
```

The `where` operator expects a data source (for example, a table) to be piped to it. The following query searches the `StormEvents` table for `DamageProperty > 0`:

```
StormEvents
| where DamageProperty > 0
```

KQL also supports compound expressions using logical operators. We will cover *arithmetic*, *logical* and *relational* operators in *Introducing KQL's scalar operators*. The following query checks for `DamageProperty > 0` in Indiana:

```
StormEvents
| where State == 'INDIANA' and DamageProperty > 0
```

As shown in *Figure 5.4*, the query returns records where the `State` is equal to `INDIANA` and the property damage is greater than `0`:

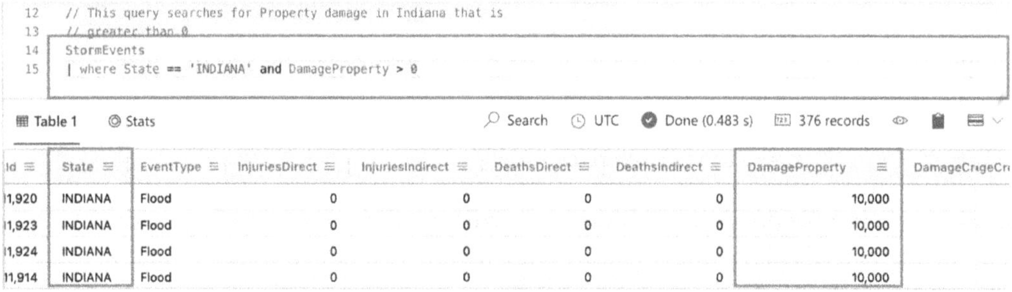

Figure 5.4 – Boolean operators used to create compound statements

The `and` operator requires both the left-hand side and the right-hand side to be `true`. If either side is `false`, the overall expression evaluates to `false`.

We will learn more about the KQL operators we can use in our expressions in *Introducing KQL's scalar operators*. Next, let's continue with the basics and look at the basic operators for aggregation in the next section.

Aggregating data and tables

In this section, we will begin to learn about *data aggregation*, and how we can perform data aggregation using the `summarize` operator. Before diving into the `summarize` operator, we should understand the `count` operator, which is commonly used with `summarize`. The `count` operator simply counts the number of rows. For example, the following query returns the total number of rows in the `StormEvents` tables:

```
StormEvents
| count
```

As shown in *Figure 5.5*, the query returns a value of 59,066. This is the number of rows in our StormEvents table:

Figure 5.5 – The count operator

Now that we understand how the count operator works, we can begin learning our data aggregation using the summarize operator.

The summarize operator

The summarize operator, which is similar to SQL's GROUP BY statement, allows us to generate aggregates based on columns. The syntax for the summarize operator is Table | summarize aggregation by column:

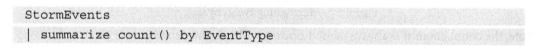

```
StormEvents
| summarize count() by EventType
```

The preceding query aggregates the EventType column, as shown in *Figure 5.6*:

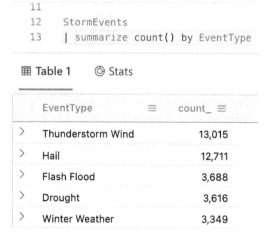

Figure 5.6 – Aggregating the EventType column

As shown in *Figure 5.6*, there are over 13,000 thunderstorms. The `summarize` operator is capable of summarizing on multiple columns. For example, the following query aggregates the number of events by state:

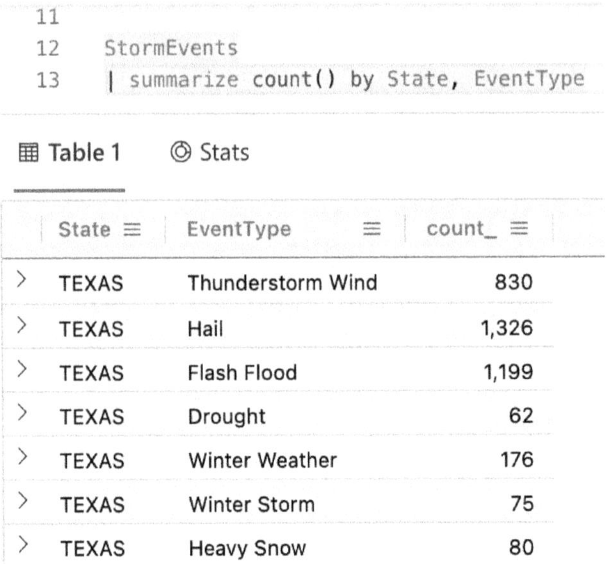

```
11
12    StormEvents
13    | summarize count() by State, EventType
```

⊞ **Table 1**   ◎ Stats

State ≡	EventType ≡	count_ ≡
> TEXAS	Thunderstorm Wind	830
> TEXAS	Hail	1,326
> TEXAS	Flash Flood	1,199
> TEXAS	Drought	62
> TEXAS	Winter Weather	176
> TEXAS	Winter Storm	75
> TEXAS	Heavy Snow	80

Figure 5.7 – Aggregating events by state

Note, the count column is always named `count_`. This is the default name given unless you explicitly specify a name. The following query shows how to change the default name of the count column to `total_events`:

```
StormEvents
| summarize total_events=count() by State, EventType
```

The next example returns the aggregated number of flash floods by state:

```
StormEvents
| where EventType == "Flash Flood"
| summarize total_incidents=count() by State, EventType
| sort by total_incidents desc
```

This query includes a `where` clause that filters our `StormEvents` table where the `EventType` is a flash flood, then aggregates the number of incidents by state.

The query then introduces the `sort` by operator, which sorts our result set by the `total_incidents` in descending order, as shown in *Figure 5.8*:

```
1    StormEvents
2    | where EventType == "Flash Flood"
3    | summarize total_incidents=count() by State, EventType
4    | sort by total_incidents desc
```

⊞ Table 1   ◎ Stats

State		EventType ≡	total_incidents ≡
>	TEXAS	Flash Flood	1,199
>	KANSAS	Flash Flood	256
>	OKLAHOMA	Flash Flood	256
>	MISSOURI	Flash Flood	251
>	IOWA	Flash Flood	223
>	ARKANSAS	Flash Flood	166
>	ARIZONA	Flash Flood	100
>	PENNSYLVANIA	Flash Flood	88

Figure 5.8 – Aggregating the number of flash floods by state

Another useful function is `bin()`. The `bin()` function allows us to bucket our values. For example, the following query shows how many updates were installed each month for the last three months:

```
Update
| where TimeGenerated >= ago(90d)
| summarize count() by bin(TimeGenerated, 30d)
| render columnchart
```

Figure 5.9 shows the security results from querying the Update table. The Update table is part of the **Log Analytics demo**, which can be accessed from https://aka.ms/lademo.

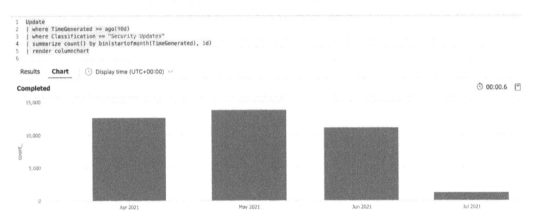

```
1  Update
2  | where TimeGenerated >= ago(90d)
3  | where Classification == "Security Updates"
4  | summarize count() by bin(startofmonth(TimeGenerated), 1d)
5  | render columnchart
6
```

Results **Chart**   ⓘ Display time (UTC+00:00) ⌄

Figure 5.9 – Security patches applied over the last 90 days

We will cover bin() in more detail in later chapters and will discuss the render statement in the next section.

Formatting output

In this section, we are going to briefly introduce the topic of *formatting output. Chapter 8, Data Visualization with Azure Data Explorer and Power BI*, covers the topic in more detail, but this serves as a good primer.

By default, when you execute a query such as the following, the result set contains all columns and you are often usually only interested in a subset of the columns:

```
StormEvents
| limit 10
```

For instance, with regard to the StormEvents tables, let's imagine we are only interested in the following columns: State, EventType, StartTime, and EndTime.

We can use the `project` operator to specify the columns we want to be included in our result set.

For example, the following query will display 10 records with the State, EventType, StartTime, and EndTime columns, as shown in *Figure 5.10*:

```
1    StormEvents
2    | project State, EventType, StartTime, EndTime
3    | limit 10
```

▦ Table 1   ◎ Stats

State	EventType ≡	StartTime	EndTime
> FLORIDA	Heavy Rain	2007-09-18 20:00:00.0000	2007-09-19 18:00:00.0000
> FLORIDA	Tornado	2007-09-20 21:57:00.0000	2007-09-20 22:05:00.0000
> ATLANTIC SOUTH	Waterspout	2007-09-29 08:11:00.0000	2007-09-29 08:11:00.0000
> AMERICAN SAMOA	Flash Flood	2007-12-07 14:00:00.0000	2007-12-08 04:00:00.0000
> KENTUCKY	Flood	2007-12-13 09:02:00.0000	2007-12-13 10:30:00.0000
> MISSISSIPPI	Thunderstor...	2007-12-20 07:50:00.0000	2007-12-20 07:53:00.0000
> MISSISSIPPI	Thunderstor...	2007-12-20 08:47:00.0000	2007-12-20 08:48:00.0000
> MISSISSIPPI	Tornado	2007-12-20 10:32:00.0000	2007-12-20 10:36:00.0000
> MISSISSIPPI	Hail	2007-12-28 02:03:00.0000	2007-12-28 02:11:00.0000
> GEORGIA	Thunderstor...	2007-12-30 16:00:00.0000	2007-12-30 16:05:00.0000

Figure 5.10 – Filtering the columns in the result set

The `project` operator also allows us to change the order of columns. For example, if we wanted to display our columns in the following order – StartTime, State, EventType, EndTime – we just specify that order when writing our query, as follows:

```
StormEvents
| project StartTime, State, EventType, EndTime
| limit 10
```

Figure 5.11 shows the columns in the result set reordered to correspond with the order used with the `project` operator:

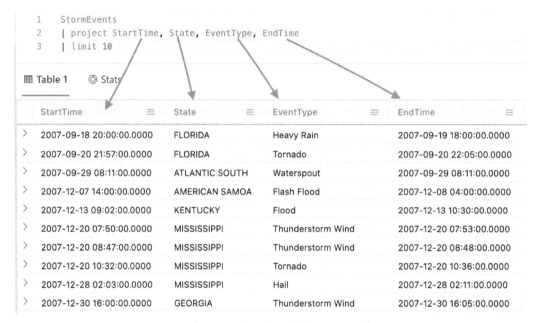

Figure 5.11 – Changing the order of the columns with project

In the next section, we will look at the `render` operator and learn how to render graphs in the ADX Web UI.

Generating graphs in the ADX Web UI

As we mentioned earlier in the chapter, a KQL query is made up of data transformation statements and an optional `render` statement to end the query. By default, result sets are returned as tables consisting of rows and columns. The `render` operator allows us to generate graphs such as bar charts and pie charts. In this section, we will briefly look at how we can generate graphs to represent our data. In *Chapter 8*, *Data Visualization with Azure Data Explorer and Power BI*, we will look at this topic in more detail and learn how to create dashboards.

Earlier in the chapter, we used the following query to aggregate all the different storm event types:

```
StormEvents
| summarize count() by EventType
```

The result set was in the form of a table consisting of two columns: the event type and the aggregated event type. With the `render` operator, we can generate a pie chart to display the same data as a graph. The following query returns the aggregated event data and renders the result as a pie chart:

```
StormEvents
| summarize total_events=count() by EventType
| render piechart
```

The pie chart is shown in *Figure 5.12*:

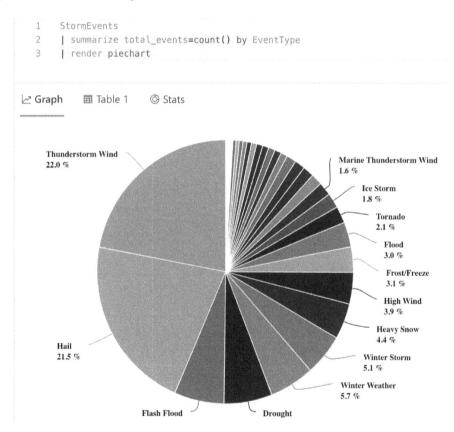

Figure 5.12 – Rendering graphs in the ADX Web UI

The `render` operator can render the following types of graphs:

- `anomalychart`
- `areachart`
- `barchart`

- card
- columnchart
- linechart
- piechart
- scatterchart
- stackedAreaChart
- table (which is the default option)
- timechart
- treemap

Figure 5.13 shows an example of a stacked area chart:

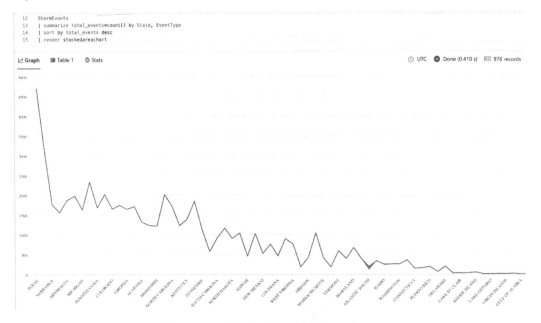

Figure 5.13 – Rendering stacked area charts

KQL provides some powerful graph rendering functionality that really brings your result sets to life. In *Chapter 8, Data Visualization with Azure Data Explorer and Power BI*, we will learn how to generate rich dashboards and reports in Power BI. In the next section, we are going to look at the explain operator that converts SQL statements to KQL queries.

Converting SQL to KQL

SQL and KQL have some similarities and most data scientists and data engineers will have some experience with SQL. KQL implements a subset of SQL and can convert your SQL statements to KQL. So, if you feel more comfortable writing in SQL, you can use the `explain` operator and KQL will generate a KQL query. For example, the following SQL statement returns all the event types for California:

```
select EventType from StormEvents where State like 'CALIFORNIA'
```

We can convert the SQL statement to KQL by prefixing it with `EXPLAIN`:

```
EXPLAIN
select EventType from StormEvents where State like 'CALIFORNIA'
```

When we run the query, the ADX Web UI will return a KQL statement, as shown in *Figure 5.14*:

Figure 5.14 – Converting SQL to KQL

In the next section, we will introduce some of the most common operators used on string, number, and date data types.

Introducing KQL's scalar operators

It is impossible to cover all of KQL's operators and functions in one chapter. But in this section, we will learn about some of the common operators. We will introduce more operators and functions throughout the rest of the book.

KQL also supports *arithmetic*, *logical*, and *relational operators*. Let's quickly review these operators, beginning with the arithmetic operators, before diving into searching and filtering, since they are commonly used in queries.

Arithmetic operators

The *arithmetic* operators allow us to perform calculations on numbers and dates. We can use the `print` operator to print strings and the result of expressions.

Table 5.4 lists the operators in order of precedence:

Operator	Description
()	Parentheses allow us to override the operator precedence.
*	Multiplies two values.
/	Divides two values.
%	Divides two numbers and returns the remainder.
+	Adds two values.
-	Subtracts two values.

Table 5.4 – Arithmetic operators for numerical values

The following `print` statement prints the result of `550 + 5`, which is `555`.

```
print 550 + 5
```

Using the parentheses, we can override the operator precedence. For example, the result of `print 5 + 5 * 100` is not the same as `print (5 + 5) * 100`. The `print 5 + 5 * 100` calculation returns `505` and `print (5 + 5) * 100` returns `1000`. This is because the parentheses takes precedence and the expression between the parentheses is calculated before the multiplication by `100`.

The arithmetic operators can also be used as part of predicates. For example, the following query returns records that have a `DamageProperty` value that is greater than or equal to `5000`. The purpose of the query is to illustrate that we can use the arithmetic operators in predicates:

```
StormEvents
| where DamageProperty >= 500 * 10
| project State, EventType, DamageProperty
```

Next, we will look at the logical operators and learn how to compare two values.

Logical operators

The *logical* operators allow us to create compound expressions by connecting two more expressions using the `and` and `or` operators.

Table 5.5 lists the logical operators with some examples, along with the rules:

Operator	Description	Examples
and	Logical and operator returns both sides of the and operator to be true.	• True and True – Evaluates to True. • True and False – Evaluates to False. • False and True – Evaluates to False. • False and False – Evaluates to False.
or	Logical or operator requires at least one side of the or operator to be true.	• True or True – Evaluates to True. • True or False – Evaluates to True. • False or True – Evaluates to True. • False or False – Evaluates to False.

Table 5.5 – Logical operators

We can use the `print` operator to test the logical operators. The `print` operator will return either `true` or `false`.

The following print statement returns `false`:

```
// returns false
print (5 + 5 == 10) and (1 + 1 == 3)
```

Figure 5.15 shows the result returned in the ADX Web UI:

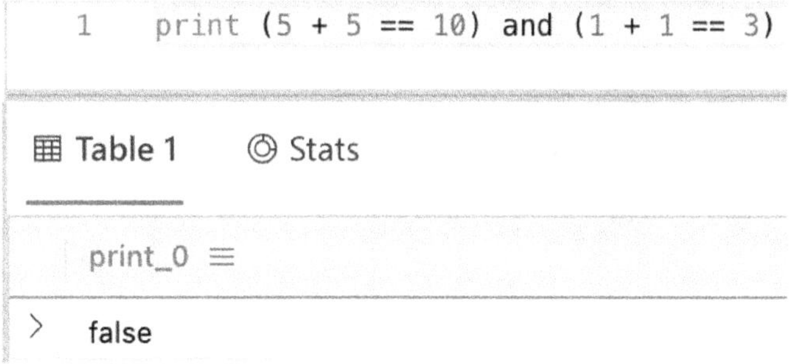

Figure 5.15 – Evaluating logical operators

The next set of operators we will look at are the relational operators, which allow us to compare values.

Relational operators

The *relational* operators allow us to compare two values. These operators are commonly used when comparing numbers and date data types

Table 5.6 shows the list of relational operators supported by KQL that we can use to compare two numeric values:

Operator	Description
<	Less than operator.
>	Greater than operator.
==	Equals operator.
!=	Not equal to operator.
<=	Less than or equal to operator.
>=	Greater than or equal to operator.
in	Checks whether a value is in a collection of numbers.
!in	Checks whether a value is not in a collection of numbers.

Table 5.6 – Relational operators for numeric values

The result of a comparison returns a Boolean value, and we can use the `print` operator to verify this. The following code snippet demonstrates the use of the `print` operator to print the result of the evaluation:

```
// returns true - 2 is less than 5
print 2 < 5


// returns false, 5 is not less than 2
print 5 < 2


// returns true, 55 is greater than 5
print 55 > 5


// returns true, 555 is equal to 555
print 555 == 555


// returns false, 555 is equal to 555
print 555 != 555


// returns true, 10 is in the sequence of numbers.
```

```
print 10 in (100, 50, 60, 70, 80, 10)
```

```
// returns true, 20 is no in the sequence of numbers.
print 20 !in (10, 30, 40, 50, 60, 70, 80)
```

The expression can also be tested in the ADX Web UI using the `print` operator. The `print` operator will print the result of an expression, `true` or `false`. *Figure 5.16* shows the result of `"Mint" == "Mint"`:

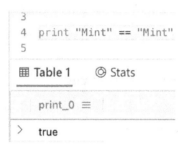

Figure 5.16 – Evaluating Boolean expressions using the print operator

As you will see in the remaining chapters, the arithmetic, logical, and relational operators are used in almost all non-trivial queries. In the next section, we are going to introduce some of the string operators that allow us to search strings.

String operators

The relational operators equals (`==`) and not equals (`!=`) can be used to compare two strings. It is important to note that the comparison is case-sensitive. There are two additional relational operators that are used to perform **case-insensitive** comparisons:

- `=~`: This checks to see whether two strings are equal but ignores the case. For example, the following `print` statement returns `true`: `print "jason" =~ "JASON"`. Whereas `print "jason" == "JASON"` returns `false`.

- `!~`: This checks whether two strings are not equal and ignores the case. For example, the following statement, `print "harrison" != "HARRISON"`, returns `true`, but `print "harrison" !~ "HARRISON"` returns `false`.

Please note, the case-sensitive operators are more efficient than the case-insensitive ones, so unless you need to perform a case-insensitive search, you should stick with the case-sensitive operators.

Next, we will look at some of the useful `string` operators. Each operator listed here is case-insensitive and each operator has a case-sensitive alternative that has the `_cs` postfix, for example, the case-sensitive version of `has` is `has_cs`.

Before looking at some of the string operators, it is important to understand the concept of a term. A *term* is a sequence of alphanumeric characters. Take the following sentence: `Azure Data Explorer is fun`. The sentence is made up of five terms: `Azure`, `Data`, `Explorer`, `is`, and `fun`.

The following list section introduces the `has` and `contains` string operators:

- `has`, `has_cs`, `!has`, and `!has_cs`: The `has` operators search for a term in a given string.

  The following `print` statement returns `true`:

  ```
  print "Azure Data Explorer is fun" has_cs "Explorer"
  ```

  But the following `print` statement returns `false`:

  ```
  print "Azure DataExplorer is fun" has_cs "Explorer"
  ```

  This is because `Explorer` is no longer a term. `DataExplorer` is the term since there is no space separating `Data` and `Explorer`.

- `Contains`, `contains_cs`, `!contains`, and `!contains_cs`: Unlike the `has` operator, the `contains` operators search for a substring.

  For instance, the following `print` statement returns `true`:

  ```
  print "Azure DataExplorer is fun" contains_cs "Explorer"
  ```

The following query queries the `SecurityDetection` table from the Log Analytics demo and searches for computers that contain the `staging` substring:

```
SecurityDetection
| where TimeGenerated >= ago(90d)
| where Computer contains_cs "staging"
```

Figure 5.17 shows the query returns one record containing a storage threat detection:

```
1   SecurityDetection
2   | where TimeGenerated >= ago(90d)
3   | where Computer contains_cs "staging"|
```

Figure 5.17 – Storage threat detection

We will introduce new string operators in the remaining chapters. Now, let's take a look at the `Date` and `Time` operators.

Date and time operators

`Date` and `Time` are two of the most common data types that you will use in KQL. KQL makes it remarkably easy to perform arithmetic operations on dates. In this section, we will cover some of the most common aspects of `Date` and `Time` and then we will jump into some examples.

> **Note**
> To demonstrate the date and time operators and functions, we will use the demo **Log Analytic** workspace Microsoft provides (https://aka.ms/LADemo) and use the `SecurityDetection` table. The `SecurityDetection` table provides information regarding the threats to your virtual machines.

The first scalar function we will look at is `ago()`. The `ago()` function can add and subtract a timespan from the current time. For example, `ago(2d)` would return the current date minus two days, such as `6/28/2021, 9:12:29.172 PM`. We can verify this using the following `print` statement:

```
print ago(2d)
```

As you will notice, the argument passed to `ago()` contains a d. This is to denote the time unit – in this case, d denotes days.

Table 5.7 lists the different postfixes we can use:

Time unit postfix	Time unit	Example
d	Days	ago(5d)
h	Hours	ago(5h)
m	Minutes	ago(45m)
s	Seconds	ago(555s)
ms	Milliseconds	ago(5ms)
microsecond	Microseconds	ago(10microsecond)
tick	nanoseconds	ago(5tick)

Table 5.7 – Time unit postfixes

When dealing with logging information and diagnostics, you will find yourself using a lot of the date/time operators and functions.

As an example, let's use the ago() function to display the security alerts that have been generated over the last five days:

```
SecurityDetection
| where TimeGenerated >= ago(5d)
| summarize count() by AlertSeverity, AlertTitle
```

As shown in *Figure 5.18*, three different types of alerts were logged over the last five days:

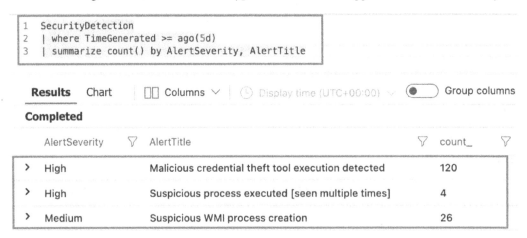

Figure 5.18 – Security alerts generated over the last five days

Another useful function is now(), which returns the current date and time. For example, print now() returns 6/30/2021, 9:54:44.449 PM.

Earlier in the chapter, we introduced the arithmetic operators – KQL allows us to use the arithmetic operators on dates and times.

The following print statement returns the number of years between two dates:

```
print (getyear(now()) - getyear(datetime(1984-10-23)))
```

In the next section, we will learn about joining tables and the different types of joins supported by KQL.

Joining tables in KQL

Like SQL, KQL supports joining tables. For those not familiar with SQL, table joins allow us to combine one or more tables in the result set. To demonstrate the use of joins, we will use two sample tables, FamilySport and FamilyFood. FamilySport contains two columns: Name, which is the name of the family member, and FavoriteSport, which is their favorite sport. The table definition is shown as follows:

```
// Create the FamilySport table that will track their favorite
sports.
.create table FamilySport (
    Name: string,
    FavoriteSport: string
)
```

FamilyFood also has two columns: Name, which is the name of the family member, and FavoriteFood, which is their favorite food. The table definition is as follows:

```
// Create the FamilyFood table that will track their favorite
foods.
.create table FamilyFood (
    Name: string,
    FavoriteFood: string
)
```

The complete table definitions, schemas, and data sources can be found at ${HOME}/Scalable-Data-Analytics-with-Azure-Data-Explorer/Chapter05/joins/.

The syntax for join queries is shown as follows:

```
Table | join [kind=jointype] on commonColumn.
```

The `join` operator joins two tables on a common column, such as `Name` in our examples. The optional `kind` argument dictates the type of comparison performed on the `commonColumn`.

The following query joins the `FamilySport` table with `FamilyFood`. The table to the left of the `join` operator is commonly referred to as the *left table* and the table to the right side of the `join` operator is referred to as the *right table*. In this example, I have not specified the `kind` of argument in order to demonstrate the default join type. The different join types will be discussed later:

```
FamilySport
| join (FamilyFood) on Name
```

The left table can be referenced as `$left` and the right table can be referenced as `$right`. The following query is equivalent to the previous query:

```
FamilySport
| join (FamilyFood) on $left.Name == $right.Name
```

Figure 5.19 shows the result of the previous two queries:

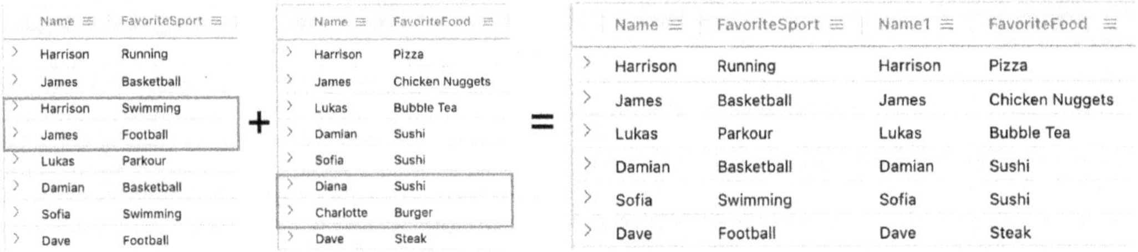

Figure 5.19 – Joining tables

The columns from `FamilySport` and `FamilyFood` have been joined where the fields in the column `Name` are equal. If you look closely at *Figure 5.19*, you will see that `Diana`, `Sushi` and `Charlotte`, `Burger` are not in the result set. This is because there was no record of Diana or Charlotte in `FamilySport`. You can also see there is no row in the result set that contains `Harrison`, `Swimming` or `James`, `Football`. This is because of the default behavior of the `join` operator in the query as we did not specify the join type in it. We will see more about this behavior by learning about the various join types.

For the remainder of this section, we will introduce the various join types in KQL:

- `innerunique`: This is KQL's default join type when you do not specify the `kind` argument. `Innerunique` joins return the first match on the common columns. In *Figure 5.19*, `Harrison, Swimming` and `James, Football` were not included in the result set because there was an earlier match for `Harrison` and `James`: `Harrison, Running` and `James, Basketball`.

- `inner`: The `inner` join is like `innerunique` joins, except it returns all matches. For instance, the following query includes `Harrison, Swimming` and `James, Football` in the result set.

```
FamilySport
| join kind=inner (FamilyFood) on Name
```

As shown in *Figure 5.20*, the query included all instances of Harrison and James in the result set:

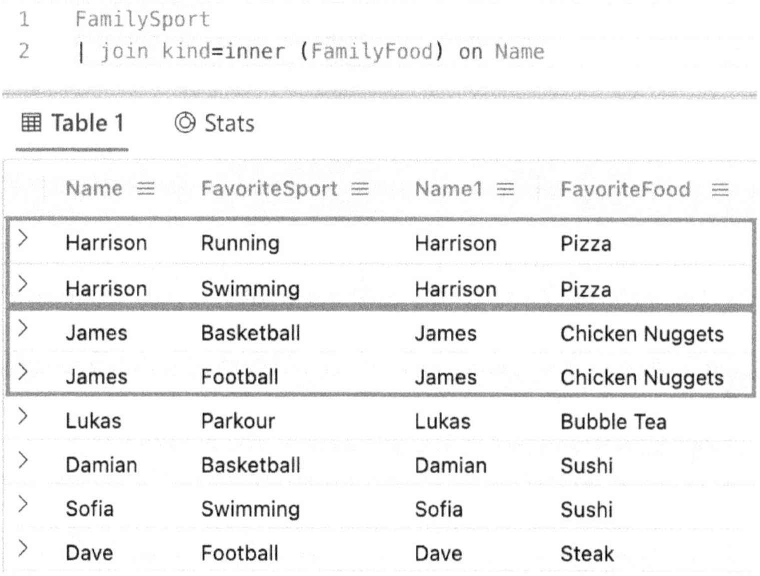

```
1    FamilySport
2    | join kind=inner (FamilyFood) on Name
```

Name ≡	FavoriteSport ≡	Name1 ≡	FavoriteFood ≡
Harrison	Running	Harrison	Pizza
Harrison	Swimming	Harrison	Pizza
James	Basketball	James	Chicken Nuggets
James	Football	James	Chicken Nuggets
Lukas	Parkour	Lukas	Bubble Tea
Damian	Basketball	Damian	Sushi
Sofia	Swimming	Sofia	Sushi
Dave	Football	Dave	Steak

Figure 5.20 – Inner join containing all matches

- **leftouter, rightouter, fullouter:** The outer family of joins returns rows, even if there is not a match. If there is no match, null (empty) values will be returned, as shown in *Figure 5.21*:

```
5    FamilySport
6    | join kind=rightouter (FamilyFood) on Name
```

⊞ **Table 1**   ◎ Stats

	Name ≡	FavoriteSport ≡	Name1 ≡	FavoriteFood ≡
>	Harrison	Running	Harrison	Pizza
>	Harrison	Swimming	Harrison	Pizza
>	James	Basketball	James	Chicken Nuggets
>	James	Football	James	Chicken Nuggets
>	Lukas	Parkour	Lukas	Bubble Tea
>	Damian	Basketball	Damian	Sushi
>	Sofia	Swimming	Sofia	Sushi
>			Diana	Sushi
>			Charlotte	Burger
>	Dave	Football	Dave	Steak

Figure 5.21 – rightouter join example

- **leftanti, rightanti:** The anti family of joins are essentially *not equals*, meaning all the rows that have no match on the common column are returned. *Figure 5.22* shows Diana and Charlotte are returned because there are no matching names in the FamilySport table. The left- / right- part of the argument denotes which table to take the results from. For example, rightanti returns rows from the right side:

```
1  ∨ FamilySport
2    | join kind=rightanti (FamilyFood) on Name
```

⊞ **FamilyFood**   ◎ Stats

	Name ≡	FavoriteFood ≡
>	Diana	Sushi
>	Charlotte	Burger

Figure 5.22 – Example of an anti join

- `leftsemi`, `rightsemi`: The `leftsemi` argument returns all the rows from the left table where there is a match on the common column and `rightsemi` returns all the rows from the right table where there is a match on the common column. As shown in *Figure 5.23*, the result set only displays the columns from the left side, which is `FamilySport` in this instance:

Figure 5.23 – Semi joins example

Each of the join types has a purpose, but in practice, you will find yourself primarily using the `inner` and `outer` joins.

In the next section, we will look at the management functions with KQL that we can use to manage our clusters, databases, and tables.

Introducing KQL's management commands

Just like SQL has its **Data Manipulation Language (DML)** and **Data Definition Language (DDL)**, KQL has them both, and they are referred to as the **KQL management commands**. The management commands allow us to manage our clusters, databases, and tables. For example, we can create, modify, and delete tables, get insights into our cluster configuration, and more. In this section, we will learn about some of the most useful management commands KQL provides. We will first look at cluster and database management commands and then we will learn about table management commands.

Cluster management

The first set of commands we will learn about are the cluster management commands. These commands provide cluster configuration and diagnostic information, which is helpful when troubleshooting issues.

The first command we will look at is `.show cluster`. The `.show cluster` management command returns a list of all nodes in the ADX cluster. As shown in *Figure 5.24*, the result set contains a record for each active node and provides detailed configuration of the nodes. This command is particularly useful when you are using scaling and want to check the size of your cluster:

Figure 5.24 – Cluster configuration information

The second cluster management command is `.show diagnostics`. The `.show diagnostics` command returns the current health state of your ADX cluster and is useful when you are troubleshooting issues such as failed data ingestion attempts.

As shown in *Figure 5.25*, the result set returns valuable information such as the number of machines that are offline and the information regarding the environment configuration:

Figure 5.25 – Cluster diagnostics

Another useful cluster management command is `.show operations`. The `.show operations` command returns a list of all the administrative commands that have been executed. This is again useful when troubleshooting issues and also for auditing.

Figure 5.26 shows a list of the administrative operations that have been performed on the cluster:

Figure 5.26 – List of administrative operations

In the next section, we will learn how to query database information and schema information, and how to manage tables.

Database and table management

The .`show databases details` command displays basic configuration information for the database, such as access control, caching policy, and sharding policy configuration.

Figure 5.27 shows the sharding policy configuration for the `samples` database:

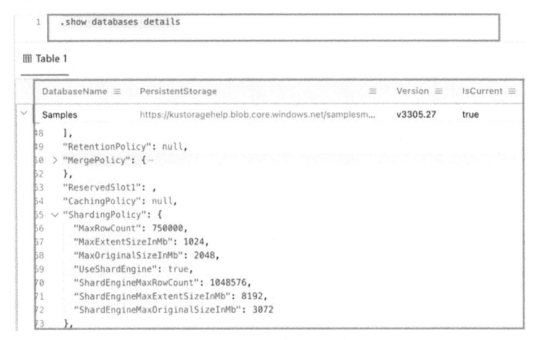

Figure 5.27 – Database configuration information

To get a list of all tables in a database, you can use the .`show tables` command, and to view table schema information, you can use .`show table <table name>`. For example, .`show table StormEvents` returns the schema for the `StormEvents` tables, as shown in *Figure 5.28*:

```
7    // returns the table schema
8    .show table StormEvents
9
```

Table 1

AttributeName	AttributeType	ExtentSize	CompressionRa...	IndexSize	IndexSizePerce...	OriginalSize	AttributeId
StartTime	DateTime	0	0	0	0	0	f5256aa3-5d0a-4aaf-99cc-3e5...
EndTime	DateTime	0	0	0	0	0	f2df9159-ed1c-420f-80d3-da14f...
EpisodeId	I32	0	0	0	0	0	41536c56-fde3-438f-9b95-96fa...
EventId	I32	0	0	0	0	0	7167ccd4-d8e2-4056-9a7b-fea...
State	StringBuffer	0	0	0	0	0	5cd21245-4a33-4ed6-92c0-66...
EventType	StringBuffer	0	0	0	0	0	d070a1b8-805c-40d6-b963-e6...
InjuriesDirect	I32	0	0	0	0	0	ac968549-627c-4d3a-9803-34...
InjuriesIndirect	I32	0	0	0	0	0	abc6001b-e397-443a-b8f1-2d5...

Figure 5.28 – StormEvents schema information

The schema information can also be returned in the **JSON** format by using the following additional arguments: `schema as json`. For example, `.show table StormEvents schema as json` returns the same schema information shown in *Figure 5.22*, just in JSON format. In *Chapter 4, Ingesting Data in Azure Data Explorer*, we learned how to create tables and schema mappings using the `.create table` command. Tables can also be deleted using the `.drop <table name>` command. For example, `.drop StormEvents` would delete our `StormEvents` table along with all its data.

Summary

This chapter is one of the most important chapters in the book in terms of reusing the skills you have learned outside of ADX clusters. As mentioned, KQL is one of the fundamental keystones to Azure with regard to managing your logging and telemetry data. Data belonging to Auditing, Security Center, Application Insights, Monitoring, and Asset Management all reside in Log Analytic workspaces, which all use KQL for querying the data.

We learned what KQL is, where it can be used, and the basic syntax of KQL queries. We then learned about the basics of KQL, such as searching, filtering with `where` clauses, aggregations with `summarize`, formatting results, rendering graphs, and converting SQL statements to KQL using the `EXPLAIN` keyword.

Next, we learned about some of the most commonly used scalar functions and operators, such as data manipulation and formatting and string search using the `has_cs` and `contains_cs` operators. We also learned how to use the `join` operator to join tables and looked at the various join types.

Then we introduced some of KQL's management commands used to manage our clusters and databases.

It's not possible to cover all the features of KQL in one chapter, but throughout the rest of this book, we will introduce new operators and functions. In the next chapter, *Chapter 6, Introducing Time Series Analysis*, we will learn about KQL's time series analysis features and in *Chapter 7, Identifying Patterns, Anomalies, and Trends in Your Data*, we will expand further on the time series analysis and learn how to detect anomalies and trends in your datasets.

Questions

Before moving on to the next chapter, test your knowledge by trying these exercises. The answers can be found at the back of the book.

1. Write a query for our EnglishPremierLeague data and aggregate the number of matches refereed by each referee.

2. What is the main difference between the search and where operators?

3. Aggregate all the event types in the StormEvents table for California and render the results as a column chart.

4. Which type of join should you use if you want to include duplicate common column matches in the result set?

6
Introducing Time Series Analysis

In the previous chapter, we introduced **Kusto Query Language** (**KQL**) and learned how to search for and filter our data using the `search` and `where` operators. Although we only scratched the surface, we introduced some of the common string, numeric, date, and time operators that help us build complex search predicates. We also saw how we can render graphs to help visualize our result sets. We can accomplish a lot with these skills in that we can review our inventory in **Azure Resource Graph** and create monitoring alerts in **Azure Monitor**.

But what if we want to analyze our data, look at the historical and current patterns, and make forecasts for the future? This is where **time series analysis** can help. In this chapter, we will learn how to convert our data into a time series.

The goal of the chapter is to remain as practical as possible, but as always, an introduction and definition of time series analysis will help to reinforce our understanding and appreciation for the power and simplicity of the time series operations and functions of KQL.

In this chapter, we will begin by introducing the concept of time series analysis, its components, and how it can help us when exploring our data. Next, we will learn how to use KQL's time series operators and functions to build a time series. We will build our time series in the **Kusto help cluster** and in the **Log Analytics demo workspace**, both of which are provided by **Microsoft**. We will also learn how to render our time series as a time chart before learning how to calculate statistics for our time series data.

In this chapter, we are going to cover the following main topics:

- What is time series analysis?
- Creating a time series with KQL
- Calculating statistics for time series data

Technical requirements

The code examples for this chapter can be found in the `Chapter06` folder of this book's GitHub repository: `https://github.com/PacktPublishing/Scalable-Data-Analytics-with-Azure-Data-Explorer.git`.

In our examples. we will be using the demo tables available on Microsoft's help cluster (`https://help.kusto.windows.net/`) and the Log Analytics playground (`https://aka.ms/lademo`), which is also provided by Microsoft.

What is time series analysis?

With our data now ingested into ADX, we can write queries using KQL to filter and search for data using search predicates. These queries are sufficient when your data is clean, structured, and you know what you are looking for, but what if we would like to do a more sophisticated exploration of our data? How can we look at trends, detect anomalies and outliers, and make forecasts, which, in turn, can help us make better decisions? One solution is time series analysis.

> **Note**
>
> The primary goal of this chapter is to focus on the practical aspects of time series analysis using the features provided by KQL. Although we will introduce the basics of time series analysis, an in-depth discussion is outside the scope of this book. There is a lot of good literature regarding time series analysis at `https://www.packtpub.com`.

Let's begin by discussing what a time series is. A time series is a sequence of data points captured over time. CPU, RAM, and disk consumption are good examples of IT-related sequences. The weather, the stock market, your heart rate, and information captured by your smartwatch are all examples of a time series.

The following diagram shows an example time series of the average hourly CPU consumption of a **virtual machine (VM)** called CH1-UBNTVM over 2 days:

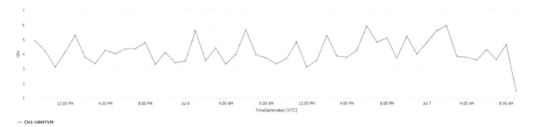

Figure 6.1 – Time series or average CPU consumption

Here, we can easily visualize the peaks and troughs (variation). Time series have four properties that we are generally interested in:

- **Trend**: This refers to the long-term direction of the data. For example, the data can have a positive growth, known as an upward trend, a negative growth known as a downward trend, and the data can also plateau.

  The following diagram shows a slight upward trend in the number of connections being made:

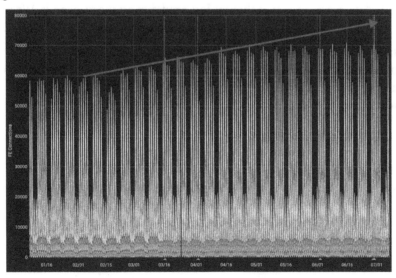

Figure 6.2 – An example of an upward trend

- **Variations**: This refers to the peaks and troughs in the data. As shown in the following diagram, the peaks and troughs are visible in the data:

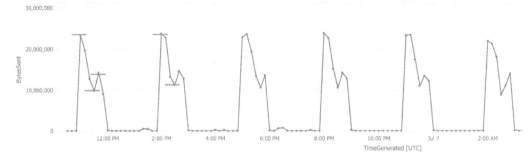

Figure 6.3 – Peaks and troughs

- **Seasonality**: This refers to reoccurring patterns at regular intervals. For instance, the following diagram illustrates the number of connections to a system:

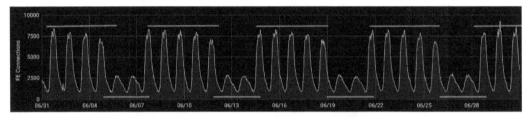

Figure 6.4 – Time series containing seasonality properties

The high peaks represent the weekdays, while the low peaks represent the weekends. From this data, we can see that there is less load during the weekends. We can use this type of information to scale our systems accordingly.

- **Cycles**: Cycles are like seasonality, meaning there is a consistent pattern, but the patterns are not consistent at regular time intervals. Referring to the preceding diagram, we can see that the weekdays reach their peaks at midday **Eastern Standard Time (EST)**.

To learn about and discover the properties of our time series, we can use an analysis method known as time series analysis to detect the properties we mentioned previously and then perform statistical analysis to identify trends, detect anomalies, and make forecasts. In the remainder of this chapter, we will learn about the operators and functions KQL provides for creating time series. In the next chapter, *Chapter 7, Identifying Patterns, Anomalies, and Trends in Your Data*, we will learn how to detect trends, anomalies, and make forecasts.

Creating a time series with KQL

KQL makes it remarkably simple to perform time series analysis and does not require you to have a background in time series analysis, although it does help. Before we dive into creating a time series, I would like to discuss some new KQL operators and functions that we will be using as helper functions.

Introducing the helper operators and functions

The first KQL feature is variables. We briefly mentioned variables in the previous chapter, *Chapter 5, Introducing the Kusto Query Language*, but it is worth spending some time learning about variables. As you will see, variables can help with the readability of your queries, which is important once you progress from the simple examples we used in the previous chapter. A few decades ago, when I first started to learn about programming, one of the first things I learned was maintainability and readability and why magic numbers are bad practice. As you will see, as we write more complex queries, values will be repeated, and rather than changing a literal value in multiple places, we can use variables.

Variables are used to store values – in the KQL context, these are variables or let statements – as they are sometimes used to allow us to assign the following:

- **Scalar types**: These are primitive types such as int, datetime, and string.
- **Tabular types**: These are the results of queries or actual tables.
- **Function body types**: These are nameless functions.

Do not worry about what the following query does; we will learn more about the make-series operator a little later, but pay close attention to the literal values we are using in the where statement ("NEW YORK") and in the data range ((datetime("2007-01-01"), datetime("2007-10-01"), 1d)):

```
StormEvents
| where StartTime between (datetime("2007-01-01") ..
datetime("2007-10-01"))
| where State == "NEW YORK"
| make-series events=count() default=0 on StartTime from
datetime("2007-01-01") to datetime("2007-10-01") step 1d by
EventType
```

At first glance, the where clause may seem obvious at this point, but what about make-series? Without understanding the make-series operator, you might make an educated guess and conclude that make-series is performing an aggregation on EventType based on StartTime.

Unless you understand what make-series is, the query is unclear. Now, consider the following query:

```
let startTime = timespan(datetime("2007-01-01"));
let endTime = timespan(datetime("2007-10-01"));
let binSize = 1d;
let nameOfState = "NEW YORK";
StormEvents
| where StartTime between (startTime .. endTime)
| where State == nameOfState
| make-series events=count() default=0 on StartTime from
startTime to endTime step binSize by EventType
```

This query is a couple of lines longer, but its readability and maintainability have been improved. Without understanding make-series, I can make an educated guess based on the variables names. This readability can be improved further by adding comments in the code.

In the next section, *Generating time series data*, where we will learn how to generate our time series data and render the data as a time chart.

Generating time series data

For the remainder of this chapter, we are going to focus on learning how to create a time series using the make-series operator, calculate statistics on our time series, render time charts, and look at some of the time series arithmetic operators and functions.

We will be using the demo tables from https://help.kusto.windows.net/, the Log Analytics playground (https://aka.ms/LADemo), and we will use a cluster to examine our StormEvents data. The help cluster contains a dataset that is perfect for demonstrating specific aspects of time series. The Log Analytics playground contains a collection of tables, which you are likely to use if you are using Azure daily.

The make-series operator creates a time series of aggregated values over a certain period. Before diving into some examples, let's examine the syntax and the arguments that make-series expects:

```
DataSource | make-series [columnName = ] aggregation [default
= default-value] on xAxisColumn [ from start] [to end] step
increment [by column]
```

The arguments in the square brackets (`[]`) are optional:

- `DataSource`: A data source needs to be piped to the operator. The data source can be a table, a nested query, or a query result that is assigned to a variable.

- `columnName =:` This is an optional name for the aggregated series. By default, the name of the column is the name of the aggregate function and its arguments if any are supplied, for example. The default name for `count()` aggregations is `count_` and the default name for `avg(param)` columns is `avg_param`.

- `aggregation`: This is a call to one of the aggregation functions and is the result of these aggregation calls that make up the *y*-axis values in the time series.

- `default = defaultValue`: This is the value that's used when there is no value available for the *y*-axis point in the specific x-bin.

- `xAxisColumn`: This is the name of the column that the series will be ordered on. For example, `date` and `time` are probably the most common columns but any numeric values can be used, such as `integers` and `floating-point numbers`.

- `from start`: This denotes the start of the time series.

- `to end`: This denotes the end of the time series.

- `step increment`: This refers to the bucket size. For example, we could have an hourly bucket size and capture the hourly average CPU consumption over a given period.

- `by column`: This specifies the column to partition the time series; for example, by user or computer. The time series can also be partitioned on multiple columns by separating them with a comma. For instance: `by EventType, State`.

Before we discuss the return value of `make-series`, let's create an example query and execute it on the help cluster. We will use the `demo_make_series1` table from the help cluster, which contains web traffic. The table contains a timestamp, the browser and operating system versions, and the country where the request originated from.

We are going to create a time series for the origin so that we can see where most of the traffic is coming from:

1. Log into the Data Explorer web UI (`https://dataexplorer.azure.com`).

2. Click on **Add Cluster**, enter `https://help.kusto.windows.net` as the connection URL, and click **Add**.

3.   In the cluster pane, expand the help cluster and click **Samples databases** to set the scope to @help/Samples. As shown in the following screenshot, your query will not execute if the scope is not set correctly:

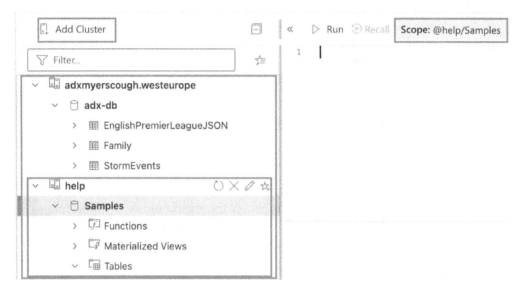

Figure 6.5 – Connecting to the help cluster

4.   Next, click **File** | **Open** and open ${HOME}/Scalable-Data-Analytics-with-Azure-Data-Explorer/Chapter06/first_time_series.kql. Click **Run**. Now, let's examine the results, as shown in the following screenshot:

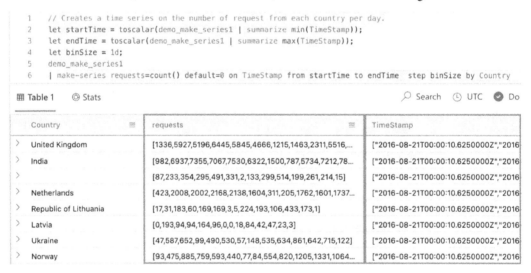

Figure 6.6 – Time series of daily web requests per country

The query returns two columns that contain arrays. The first is `requests`, which is an array of the total daily requests and an array of dates. You can think of the `requests` array as our *y*-axis values and the `TimeStamp` array as the *x*-axis values. The first element in the results array is bound to the first element of `TimeStamp` and so on.

The following diagram illustrates this mapping:

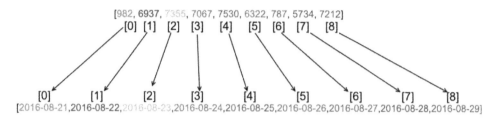

Figure 6.7 – Mapping the y-axis values to the x-axis values

Now, let's look at the query that was used to generate the time series and break it down line by line:

```
let startTime = toscalar(demo_make_series1 | summarize
min(TimeStamp));

let endTime = toscalar(demo_make_series1 | summarize
max(TimeStamp));

let binSize = 1d;

demo_make_series1

| make-series requests=count() default=0 on TimeStamp from
startTime to endTime step binSize by Country
```

As promised in the previous chapter, we will be introducing more and more operators and functions throughout this book. This query introduces the `toscalar()` function and the `between` operator.

Now, let's step through the query line by line to understand what is happening:

1. `let startTime = toscalar(demo_make_series1 | summarize min(TimeStamp));` returns the earliest date available in the dataset by using the `min()` aggregation function and assigns the result to the `startTime` variable. The `toScalar()` function returns a constant value of the nested `summarize` query.

2. `let endTime = toscalar(demo_make_series1 | summarize max(TimeStamp));` is similar to *line 1* except it assigns the last date in the dataset and assigns the value to `endTime`.

3. `let binSize = 1d;` assigns the bin size to the `binSize` variable. In this case, we want a bin size of 1 day, which means we are capturing the total number of requests per day.

4.   `demo_make_series1` is the data source we are querying.

5.   `| make-series requests=count() default=0 on TimeStamp from startTime to endTime step binSize by Country` generates our aggregated time series by counting the number of daily (`binSize`) requests by country.

One of the best features of the `make-series` operator is that it can be piped to the `render` statement, and `render timechart` knows which values to use to generate a time chart. The following query is an updated version of the previous query containing `render timechart` so that we can generate a time chart of our time series:

```
1    let startTime = toscalar(demo_make_series1 | summarize min(TimeStamp));
2    let endTime = toscalar(demo_make_series1 | summarize max(TimeStamp));
3    let binSize = 1d;
4    demo_make_series1
5    | make-series requests=count() default=0 on TimeStamp from startTime to endTime step binSize by Country
6    | render timechart
```

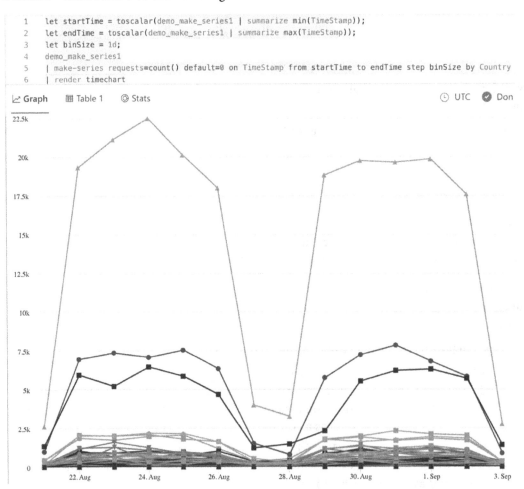

Figure 6.8 – Rendering a time series as a time chart

Here, we can immediately begin to see patterns. We will learn about pattern detection in *Chapter 7, Identifying Patterns, Anomalies, and Trends in Your Data*, but here, we can see that traffic is low on the weekend compared to the weekdays and that most of the traffic is coming from the US.

As you know, make-series returns arrays of the aggregated time series data and the *x* axis. There is an operator that allows us to split our arrays into records called mv-expand.

The following query creates a time series of the total number of requests and uses mv-expand to split the arrays into rows in our table:

```
let startTime = toscalar(demo_make_series1 | summarize min(TimeStamp));
let endTime = toscalar(demo_make_series1 | summarize max(TimeStamp));
let binSize = 1d;
demo_make_series1
| make-series requests=count() default=0 on TimeStamp from startTime to endTime step binSize by Country
| mv-expand TimeStamp to typeof(datetime), requests to typeof(long)
| order by Country
```

The important line in the aforementioned query is mv-expand TimeStamp to typeof(datetime), requests to typeof(int). The mv-expand operator splits each array into scalar values of a specified data type – in this case, datetime and integers.

The result of the query is shown in the following screenshot:

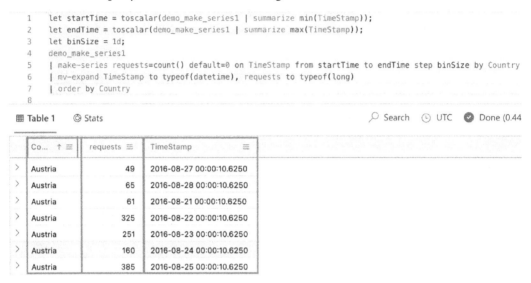

```
1    let startTime = toscalar(demo_make_series1 | summarize min(TimeStamp));
2    let endTime = toscalar(demo_make_series1 | summarize max(TimeStamp));
3    let binSize = 1d;
4    demo_make_series1
5    | make-series requests=count() default=0 on TimeStamp from startTime to endTime step binSize by Country
6    | mv-expand TimeStamp to typeof(datetime), requests to typeof(long)
7    | order by Country
8
```

Co... ↑ ≡	requests ≡	TimeStamp ≡
Austria	49	2016-08-27 00:00:10.6250
Austria	65	2016-08-28 00:00:10.6250
Austria	61	2016-08-21 00:00:10.6250
Austria	325	2016-08-22 00:00:10.6250
Austria	251	2016-08-23 00:00:10.6250
Austria	160	2016-08-24 00:00:10.6250
Austria	385	2016-08-25 00:00:10.6250

Figure 6.9 – Splitting arrays into records

Let's consider another scenario. The Log Analytics playground provided by Microsoft allows us to get hands-on experience with common Azure telemetry tables such as `Perf` for performance counters and `SecurityEvents` for security-related issues. Another useful table is `Update`, which provides us with data regarding operating system patches. We are going to build a query to generate a time series that shows how many security patches have been applied weekly for the past 100 days.

The following query should look familiar by now:

```
let startTime = ago(100d);
let endTime = now();
let binSize = 7d;
Update
| where Classification == "Security Updates"
| make-series security_updates=count() default=0 on
TimeGenerated from startTime to endTime step binSize by
UpdateState
| render timechart
```

This query is almost identical to the one we used earlier to query the demo_make_series1 table. The main differences are the table and column names. This emphasizes the power and simplicity of KQL and the value of variables in terms of readability and maintainability.

Next, let's step through the query line by line to understand what is happening:

1. `let startTime = ago(100d);` assigns the date of 100 days before assigning the current date and time to `startTime`.

2. `let endTime = now();` assigns the current date and time to `endTime`.

3. `let binSize = 7d;` sets our bin size for our aggregations to weekly.

4. `Update` is the operating system patching information table.

5. `| where Classification == "Security Updates"` filters the query based on whether it has been classified as a security update.

6. `| make-series security_updates=count() default=0 on TimeGenerated from startTime to endTime step binSize by UpdateState` generates a time series that is partitioned on `UpdateStated`. The value of `UpdateState` is either installed or needed.

7. `| render timechart` generates a time chart showing the number of patches that have been installed and that were still needed weekly for the last 100 days.

To execute the query, follow these steps:

1. Go to `https://aka.ms/LADemo`. You will be redirected to the Azure portal. If you are prompted for credentials, enter the credentials you use to log into Azure.

2. Copy and paste the query into the Query Editor and press *Shift + Enter* to execute it. Note that the **Log Analytics Query Editor** is different from the **Data Explorer Web UI** and that there are limitations regarding the KQL commands you can run; for instance, the management commands will not work.

The following screenshot shows the results of the query. As you can see, at the time of writing, there is a downward trend. Please note that when you execute the query, the results may change since this data is updated daily:

```
1   // Generates a time series for the number of security updates installed in the last 100 days.
2   let startTime = ago(100d);
3   let endTime = now();
4   let binSize = 7d;
5   Update
6   | where Classification == "Security Updates"
7   | make-series security_updates=count() default=0 on TimeGenerated from startTime to endTime step binSize by UpdateState
8   | render timechart
9
```

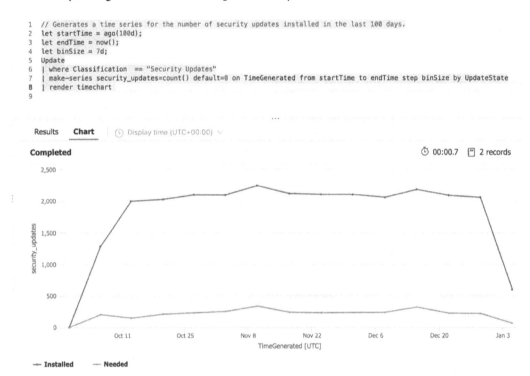

Figure 6.10 – Time series showing the number of security patches that have been installed

Before wrapping up this chapter, there is one more function I would like to cover and that is `series_stats()`. We will look at it in the next section, *Calculating statistics for time series data*.

Calculating statistics for time series data

Another useful function is `series_stats()`. The `series_stats()` function takes one or multiple time series as an argument and returns the following statistical information:

- `min`: The minimum value in the time series.
- `min_idx`: The index of the minimum value in the time series.
- `max`: The maximum value in the time series.
- `max_idx`: The index of the maximum value in the time series.

- avg: The average value of the time series.

- variance: The sample variance of the time series. The sample variance is the squared deviation of the time series's mean. The sample variance is used to calculate the standard deviation.

- stdev: The sample standard deviation. The standard deviation is the amount of variation in the values of the time series.

The following screenshot shows the statistics for our security patching time series:

Figure 6.11 – Calculating statistics for a time series

As you can see, the series_stats() function appends multiple columns to our result set.

Summary

Congratulations on completing this chapter! In this chapter, we started by introducing time series and time series analysis. You learned what a time series is and the properties of a time series, such as trends, seasonality, variations, and cycles.

Next, we dived into the practical aspects of time series and learned how to create them using the make-series operator. You learned that the make-series operator returns the series as arrays and that the time chart can render this data without any additional information.

We then worked through a couple of examples and generated a time series. We created a time series in the help cluster and generated a time chart to visualize the number of requests per country. We then learned about the Log Analytics demo workspace and generated a time series to understand how many security patches had been applied in the last 100 days.

Finally, we looked at the `series_stats()` function, which calculates statistics for our time series data.

In the next chapter, *Chapter 7, Identifying Patterns, Anomalies, and Trends in Your Data*, we will expand on time series analysis and learn how to detect anomalies, trends, and forecasts in datasets.

Questions

Before moving on to the next chapter, test your knowledge by answering these questions. The answers can be found at the back of this book:

1. What are the properties of a time series?

2. Which operator can we use to generate a time series?

3. Can you fill in the blanks of this query to display the number of patches that have been installed in the last 30 days and render the results as a time chart?

```
let startTime = ago(_____);
let endTime = now();
let binSize = 7d;
Update
| where Classification == "Security Updates"
| make-series security_updates=count() default=0 on
TimeGenerated from startTime to endTime step _____ by
UpdateState
| render _____
```

4. Using `mv-expand`, split the following time series into records:

```
let startTime = ago(100d);
let endTime = now();
let binSize = 7d;
Update
| where Classification == "Security Updates"
| make-series security_updates=count() default=0 on
TimeGenerated from startTime to endTime step binSize by
UpdateState
```

7
Identifying Patterns, Anomalies, and Trends in your Data

In the previous chapter, we introduced the concept and properties of **time series** and demonstrated how to create time series and render them as **time charts** in **Kusto Query Language (KQL)**. Now that we are familiar with time series and their properties such as **seasonality**, **variations**, and **trends**, the next step is to learn how to identify these patterns and properties in our data.

The goal of the chapter is to remain as practical as possible and focus on learning how and when to use KQL's functions and operators, which allow us to analyze our data, identify trends and anomalies, and make forecasts so that we can gain better insights into our data.

In this chapter, we will begin by learning about **moving averages** and how moving averages can help reduce noise and smoothen our time series data. Next, we will learn how to perform **trend analysis** in KQL by using **linear regression**.

Finally, we will learn how to determine the **seasonality**, **trend**, and **residual components** of our time series using `series_decompose()`. Then, we will learn how we can use these components to detect anomalies and calculate forecasts.

In this chapter, we are going to cover the following main topics:

- Calculating moving averages with KQL
- Trend analysis with KQL
- Anomaly detection and forecasting with KQL

Technical requirements

The code examples for this chapter can be found in the `Chapter07` folder of this book's GitHub repository: `https://github.com/PacktPublishing/Scalable-Data-Analytics-with-Azure-Data-Explorer`.

In our examples, we will be using the demo tables that are available on the **Microsoft help cluster** (`https://help.kusto.windows.net/`) and the **Log Analytics playground** (`https://aka.ms/lademo/`), which is also provided by Microsoft.

Calculating moving averages with KQL

There may be instances where your time series data is clean and all the components such as seasonality, trends, and variations are visible to the point that you can confidently make decisions based on the data without having to manipulate or clean the data. In reality, there will be noise and variations in the data that may obfuscate patterns and anomalies. KQL provides a rich set of functions for analyzing time series data. One subset of those functions is for calculating moving averages. Moving averages allow us to remove noise and smoothen our data.

The goal of this section is to learn how to use `series_fir()` to calculate moving averages to smoothen our data. **Finite Impulse Response** (**FIR**) is a filtering technique that is commonly used in signal processing and time series.

As you may recall, in *Chapter 6, Introducing Time Series Analysis*, we used `demo_make_series1`, which is a table in the help cluster (`https://help.kusto.windows.net`) to render the number of `HTTP` requests per country. As shown in the following diagram, most requests came from the United States:

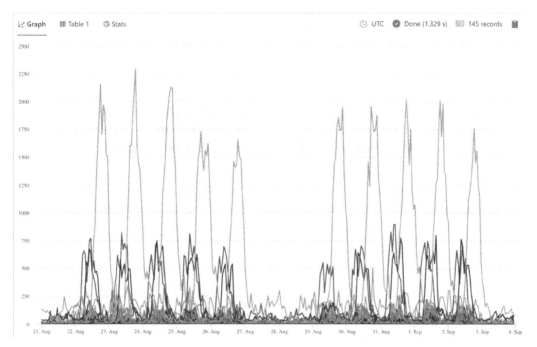

Figure 7.1 – HTTP requests per country

The preceding diagram also illustrates clear seasonality in the dataset, with most traffic occurring during the week.

Before using `series_fir()` and calculating our moving averages, I would like to introduce one of the new helper functions we will be using in this chapter. The `repeat(number, count)` function generates a dynamic array containing `number`, `count` times.

For instance, the following example generates a dynamic array containing three instances of five:

```
print repeat(5, 3)
```

The output of the print statement is `[5, 5, 5]`. We will be using the `repeat` function when we define our filter size.

In the next example, we will calculate a moving average using `series_fir()` to smoothen our data. The `series_fir(x, filter, normalize, center)` function takes four arguments:

- `x` is the column we want to calculate the moving average on.

- `filter` stores the coefficients of the filter. As we will see, the larger the filter, the smoother we can make our curves. Without going into too much theory, the coefficients are the values used by the FIR algorithm to filter/smoothen our time series.

- `normalize` is a Boolean argument that determines whether the filter's coefficients are normalized, meaning the sum of the coefficients is equal to one. Normalizing the filter ensures our time series is not amplified or attenuated.

  Let's imagine we have a filter containing the following values: `[1, 1, 1, 1, 1]`. When we normalize the filter, the values become `[0.20, 0.20, 0.20, 0.20, 0.20]`.

- `center` is a Boolean value that determines whether the filter is symmetrical.

  For streaming data we should use `center=false`, as we do not have future values after the current point. Therefore, we can average only from `v=current` point backwards.

  For analyzing existing data, it is recommended to use `center=true` so the moving average of specific sample will include both preceding and successive samples.

  By default, `center` is set to `false`. When set to `true`, the current data point in our time series is the middle value in the filter. For instance, let's assume we have a time series with the following values: `ts = [1, 2, 3, 4, 5]`.

  When the value of `center` is false, then the elements are added to the result array from right to left. Our filter for each element would be as follows:

  - `f[0] = [0, 0, 0, 0, 1] = 0.2`
  - `f[1] = [0, 0, 0, 1, 2] = 0.6`
  - `f[2] = [0, 0, 1, 2, 3] = 1.2`
  - `f[3] = [0, 1, 2, 3, 4] = 2`
  - `f[4] = [1, 2, 3, 4, 5] = 3`

You can imagine the time series (ts) flowing into the filter (f) from right to left, as shown in the following diagram:

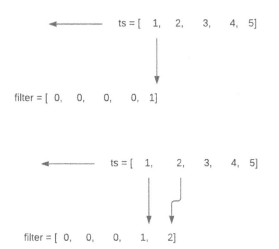

Figure 7.2 – FIR filtering when center is set to false

If we set center to true, then our elements of ts will be added to the center of our filter (f). The result would be as follows:

- f[0] = [0, 0, 1, 2, 3] = 1.2
- f[1] = [0, 1, 2, 3, 4] = 2
- f[2] = [1, 2, 3, 4, 5] = 7.5
- f[3] = [2, 3, 4, 5, 0] = 2.8
- f[4] = [3, 4, 5, 0, 0] = 2.4

The following query, which can be executed against the help cluster in the ADX Web UI, generates a moving average for the number of HTTP requests made from United States:

```
let startTime = toscalar(demo_make_series1 | summarize
min(TimeStamp));
let endTime = toscalar(demo_make_series1 | summarize
max(TimeStamp));
let binSize = 1h;
let windowSize = 10;
demo_make_series1
```

```
| where Country == "United States"
| make-series requests=count() default=0 on TimeStamp from
startTime to endTime step binSize by Country
| extend movingAverage = series_fir(requests, repeat(1,
windowSize), true, true)
| render timechart
```

The output of the preceding query is as follows:

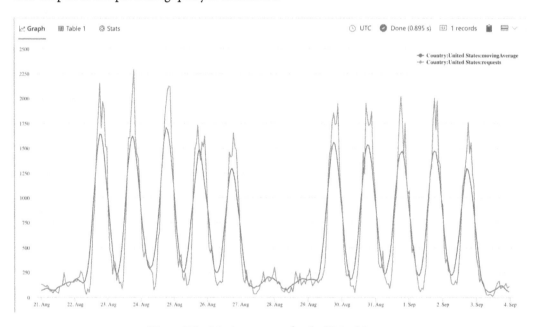

Figure 7.3 – Moving average for the United States

The preceding diagram depicts two time series on the time chart time. The time series with the highest peaks is our original series, while the smaller, smoother series is our moving average. As you can see, the smoother data series removes the spikes (variations), making it easier to identify the seasonality component.

Now, let's break down the previous query line by line:

1. `let startTime = toscalar(demo_make_series1 | summarize min(TimeStamp));` returns the earliest date available in the dataset by using the `min()` aggregation function and assigns the result to the `startTime` variable. The `toscalar()` function returns a constant value of the nested `summarize` query.

2. `let endTime = toscalar(demo_make_series1 | summarize max(TimeStamp));` is similar to *line 1* except it assigns the last date in the dataset and assigns the value to `endTime`.

3. `let binSize = 1h;` assigns the bin size to the `binSize` variable. In this case, we want a bin size of 1 day, which means we are capturing the total number of requests per day.

4. `let windowSize = 10;` is the size of the filter we will use to calculate the moving average. Since we are normalizing the coefficients, the values of the filter will be `[0.1, 0.1, 0.1, 0.1, 0.1, 0.1, 0.1, 0.1, 0.1, 0.1]`.

5. `demo_make_series1` is the data source we are querying.

6. `| where Country == "United States"` filters the country to just include United States.

7. `| make-series requests=count() default=0 on TimeStamp from startTime to endTime step binSize by Country` generates our aggregated time series by counting the number of daily (`binSize`) requests by country.

8. `| extend movingAverage = series_fir(requests, repeat(1, windowSize), true, true)` calculates our moving average for the time series we generated in *line 8*. We are also normalizing and centering the moving average.

9. `| render timechart` generates the time chart by showing the number of HTTP requests from the United States and the moving average.

KQL provides a helper user-defined function that does the same as `series_fir()`; it just has a more readable name: `series_moving_avg_fl()`. The following query generates the same output as our previous query. As you can see, `| extend movingAverage = series_moving_avg_fl(requests, windowSize, true)` is more readable than *line 9* of our original query, which was discussed previously.

Under the hood, `series_moving_avg_fl()` simply calls `series_fir()`. The following code snippet is the implementation of `series_moving_avg_fl()`:

```
let series_moving_avg_fl = (y_series:dynamic, n:int,
center:bool=false)
{
    series_fir(y_series, repeat(1, n), true, center)
}
```

As you can see, `series_moving_avg_fl()` always normalizes the coefficients by setting the third argument for `series_fir()` to `true` and takes care of calling `repeat()` to create the filter:

```
let series_moving_avg_fl = (y_series:dynamic, n:int,
center:bool=false)
{
    series_fir(y_series, repeat(1, n), true, center)
}
```

With regards to **Azure Log Analytics**, there are some subtle differences that you need to be aware of. You can add it by saving the previous code snippet as a function. Let's create a new **Log Analytics workspace** to demonstrate how to save user-defined functions.

The following sequence of steps demonstrates how to deploy a new Log Analytics workspace using our ARM templates from **Azure Cloud Shell**:

1. Open Cloud Shell by going to `https://shell.azure.com`.

2. If you have not cloned the repository, run the `git clone https://github.com/PacktPublishing/Scalable-Data-Analytics-with-Azure-Data-Explorer.git` command.

3. Navigate to the source directory by typing `cd Scalable-Data-Analytics-with-Azure-Data-Explorer/Chapter07`.

4. Next, deploy the Log Analytics workspace into our `adx-rg` resource group by typing `New-AzResourceGroupDeployment -Name "LogAnalyticsDeployment" -ResourceGroup "adx-rg" -TemplateFile ./logAnalyticsWorkspace.json -TemplateParameterFile ./logAnalyticsWorkspace.params.json`. Like all Azure resources, the deployment can take a couple of minutes, depending on how busy the platform is at a given point in time.

Now that we have deployed our Log Analytics workspace, the next step is to deploy to save `series_moving_avg_fl()` as a function. The following sequence of steps explains how to save `series_moving_avg_fl()` as a function in our Log Analytics workspace so that it can be used by our other queries:

1. Sign into the Azure portal (`https://portal.azure.com`).

2. Search for `PacktLA` in the search bar at the top of the portal and click on the workspace, as shown in the following screenshot:

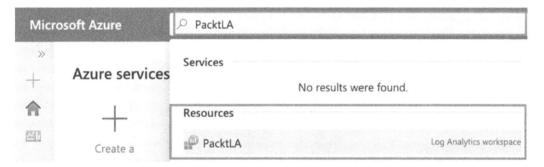

Figure 7.4 – Searching for the PacktLA workspace

3. Under the workspace's general properties, click **Logs**. You will be prompted by the queries dialog, which displays some example queries. Close the dialog and paste our user-defined function into the query editor.

4. Click **Save**, then **Save as function**. Populate the function's properties and click **Save**, as shown in the following screenshot:

Figure 7.5 – Saving the helper function to our workspace

The function will now be available in your workspace, and you can use it as part of your queries.

The second subtle difference between Azure Data Explorer Web UI and Log Analytics workspace is the way the `render` operator works. In the ADX Web UI, the `render` operator can work directly with dynamic types. With Log Analytics, we must expand our time series using `mv-expand` and convert our values from dynamic into their specific data types. Then, we must project the columns we want to pipe to the `render` operator.

The following query generates a time series and moving average for the average CPU consumption for the computer named `CH1-UBNTVM`:

```
let startTime = ago(1d);
let endTime = now();
let binSize = 10m;
let coefficient_windowsize=5;
Perf
| where TimeGenerated >= startTime
| where CounterName == "% Processor Time" and Computer == "CH1-UBNTVM"
| make-series cpu=avg(CounterValue) default=0 on TimeGenerated from startTime to endTime step binSize by Computer
| extend movingAverage=series_fir(cpu, repeat(1, coefficient_windowsize))
```

Since the `Perf` data is constantly updated by Microsoft, feel free to adjust the `startTime` variable to an earlier time. For example, `30d` would create a window of the last 30 days.

In the previous line, we calculate our moving average by calling `series_fir()`:

```
| mv-expand cpu to typeof(double), movingAverage to typeof(double), TimeGenerated to typeof(datetime)
| render timechart
```

Line 16 then renders both time series by piping the data to `render`.

The output of the previous query is as follows:

```
 6   Let startTime = ago(1d);
 7   let endTime = now();
 8   let binSize = 10m;
 9   let coefficient_windowsize=5;
10   Perf
11   | where TimeGenerated >= startTime
12   | where CounterName == "% Processor Time" and Computer == "CH1-UBNTVM"
13   | make-series cpu=avg(CounterValue) default=0 on TimeGenerated from startTime to endTime step binSize
14   | extend movingAverage=series_fir(cpu, repeat(1, coefficient_windowsize))
15   | mv-expand cpu to typeof(double), movingAverage to typeof(double), TimeGenerated to typeof(datetime)
16   | render timechart
```

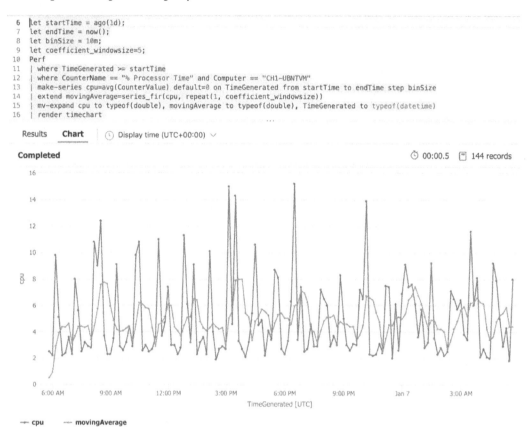

Figure 7.6 – Rendering a moving average in Log Analytics

In the preceding screenshot, *line 10*, mv-expand cpu to typeof(double), movingAverage to typeof(double), TimeGenerated to typeof(datetime), expands the dynamic array of moving averages and converts the data types into their concrete data types. For example, the CPU values are converted into doubles.

In this section, you were introduced to moving averages, which can be used to remove noise from your time series. This can help you identify the components of time series data such as variations, seasonality, and so on. Reducing noise from our time series can also improve accuracy when we want to perform other operations such as anomaly detection and forecasting, which we will cover later in this chapter. In the next section, we will learn how to perform linear regression on time series data to gain insights into trends within our time series.

The second subtle difference between Azure Data Explorer Web UI and Log Analytics workspace is the way the render operator works. In the ADX Web UI, the render operator can work directly with dynamic types. With Log Analytics, we must expand our time series using mv-expand and convert our values from dynamic into their specific data types. Then, we must project the columns we want to pipe to the render operator.

The following query generates a time series and moving average for the average CPU consumption for the computer named CH1-UBNTVM:

```
let startTime = ago(1d);
let endTime = now();
let binSize = 10m;
let coefficient_windowsize=5;
Perf
| where TimeGenerated >= startTime
| where CounterName == "% Processor Time" and Computer == "CH1-
UBNTVM"
| make-series cpu=avg(CounterValue) default=0 on TimeGenerated
from startTime to endTime step binSize by Computer
| extend movingAverage=series_fir(cpu, repeat(1, coefficient_
windowsize))
```

Since the Perf data is constantly updated by Microsoft, feel free to adjust the startTime variable to an earlier time. For example, 30d would create a window of the last 30 days.

In the previous line, we calculate our moving average by calling series_fir():

```
| mv-expand cpu to typeof(double), movingAverage to
typeof(double), TimeGenerated to typeof(datetime)
| render timechart
```

Line 16 then renders both time series by piping the data to render.

Since num is already a time series, we do not need to call make-series. We can simply call series_fit_line() and pipe the result to render timechart. The following query, which can be executed against the help cluster in https://dataexplorer.azure.com, plots the best fit line on our time chart:

```
demo_series3
| extend
(RSquare,Slope,Variance,RVariance,Interception,LineFit)=series_
fit_line(num)
| render timechart
```

The output of the preceding query is as follows:

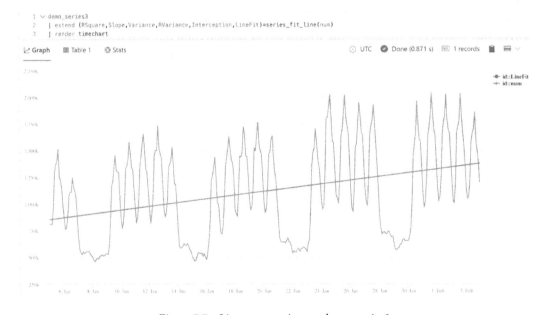

Figure 7.7 – Linear regression on demo_series3

As you can see, the straight line crossing the graph (id::LineFit), which is our best fit line, is showing an upward trend, which indicates that the number of requests is increasing. We can also look at the other values returned by series_fit_line() by projecting the values in tabular form.

The second subtle difference between Azure Data Explorer Web UI and Log Analytics workspace is the way the render operator works. In the ADX Web UI, the render operator can work directly with dynamic types. With Log Analytics, we must expand our time series using mv-expand and convert our values from dynamic into their specific data types. Then, we must project the columns we want to pipe to the render operator.

The following query generates a time series and moving average for the average CPU consumption for the computer named CH1-UBNTVM:

```
let startTime = ago(1d);
let endTime = now();
let binSize = 10m;
let coefficient_windowsize=5;
Perf
| where TimeGenerated >= startTime
| where CounterName == "% Processor Time" and Computer == "CH1-
UBNTVM"
| make-series cpu=avg(CounterValue) default=0 on TimeGenerated
from startTime to endTime step binSize by Computer
| extend movingAverage=series_fir(cpu, repeat(1, coefficient_
windowsize))
```

Since the Perf data is constantly updated by Microsoft, feel free to adjust the startTime variable to an earlier time. For example, 30d would create a window of the last 30 days.

In the previous line, we calculate our moving average by calling series_fir():

```
| mv-expand cpu to typeof(double), movingAverage to
typeof(double), TimeGenerated to typeof(datetime)
| render timechart
```

Line 16 then renders both time series by piping the data to render.

Since num is already a time series, we do not need to call make-series. We can simply call `series_fit_line()` and pipe the result to `render timechart`. The following query, which can be executed against the help cluster in `https://dataexplorer.azure.com`, plots the best fit line on our time chart:

```
demo_series3
| extend
(RSquare,Slope,Variance,RVariance,Interception,LineFit)=series_
fit_line(num)
| render timechart
```

The output of the preceding query is as follows:

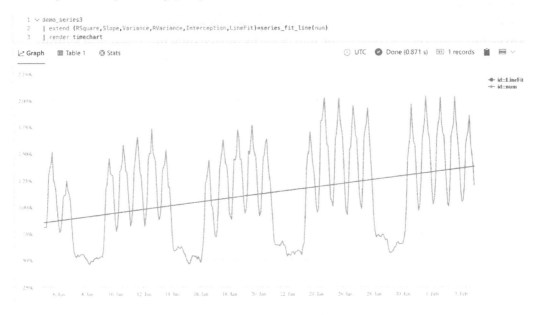

Figure 7.7 – Linear regression on demo_series3

As you can see, the straight line crossing the graph (`id::LineFit`), which is our best fit line, is showing an upward trend, which indicates that the number of requests is increasing. We can also look at the other values returned by `series_fit_line()` by projecting the values in tabular form.

The second subtle difference between Azure Data Explorer Web UI and Log Analytics workspace is the way the render operator works. In the ADX Web UI, the render operator can work directly with dynamic types. With Log Analytics, we must expand our time series using mv-expand and convert our values from dynamic into their specific data types. Then, we must project the columns we want to pipe to the render operator.

The following query generates a time series and moving average for the average CPU consumption for the computer named CH1-UBNTVM:

```
let startTime = ago(1d);
let endTime = now();
let binSize = 10m;
let coefficient_windowsize=5;
Perf
| where TimeGenerated >= startTime
| where CounterName == "% Processor Time" and Computer == "CH1-UBNTVM"
| make-series cpu=avg(CounterValue) default=0 on TimeGenerated from startTime to endTime step binSize by Computer
| extend movingAverage=series_fir(cpu, repeat(1, coefficient_windowsize))
```

Since the Perf data is constantly updated by Microsoft, feel free to adjust the startTime variable to an earlier time. For example, 30d would create a window of the last 30 days.

In the previous line, we calculate our moving average by calling series_fir():

```
| mv-expand cpu to typeof(double), movingAverage to typeof(double), TimeGenerated to typeof(datetime)
| render timechart
```

Line 16 then renders both time series by piping the data to render.

The output of the previous query is as follows:

```
 6  let startTime = ago(1d);
 7  let endTime = now();
 8  let binSize = 10m;
 9  let coefficient_windowsize=5;
10  Perf
11  | where TimeGenerated >= startTime
12  | where CounterName == "% Processor Time" and Computer == "CH1-UBNTVM"
13  | make-series cpu=avg(CounterValue) default=0 on TimeGenerated from startTime to endTime step binSize
14  | extend movingAverage=series_fir(cpu, repeat(1, coefficient_windowsize))
15  | mv-expand cpu to typeof(double), movingAverage to typeof(double), TimeGenerated to typeof(datetime)
16  | render timechart
```

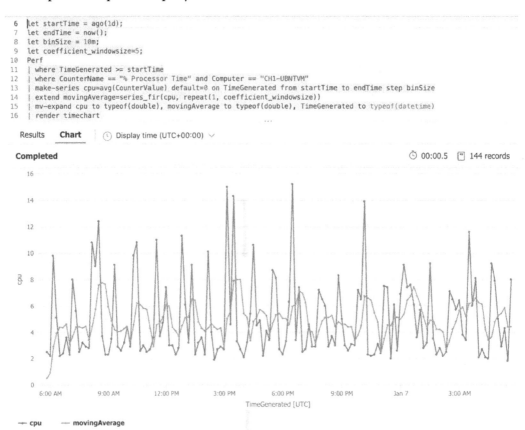

Figure 7.6 – Rendering a moving average in Log Analytics

In the preceding screenshot, *line 10*, `mv-expand cpu to typeof(double)`, `movingAverage to typeof(double)`, `TimeGenerated to typeof(datetime)`, expands the dynamic array of moving averages and converts the data types into their concrete data types. For example, the CPU values are converted into `doubles`.

In this section, you were introduced to moving averages, which can be used to remove noise from your time series. This can help you identify the components of time series data such as variations, seasonality, and so on. Reducing noise from our time series can also improve accuracy when we want to perform other operations such as anomaly detection and forecasting, which we will cover later in this chapter. In the next section, we will learn how to perform linear regression on time series data to gain insights into trends within our time series.

Trend analysis with KQL

As we discussed in *Chapter 6*, *Introducing Time Series Analysis*, one of the components of time series data is a trend. A **trend** helps visualize and predict the long-term direction of data. The trend is either positive, also known as an **upward trend**, or negative, also known as a **downward trend**. KQL provides two functions, `series_fit_line()` and `series_fit_2lines()`, for calculating the trend. We will begin by looking at `series_fit_line()` before looking at `series_fit_2lines()`.

Applying linear regression with KQL

The `series_fit_line()` function performs linear regression to calculate the best fit line, also known as the **regression line**, for our original time series. Once we have calculated our regression line, we can identify the positive and negative relationships between our x-axis, also known as the independent variable, and our y-axis, also known as the dependent variable. The `series_fit_line()` function takes one argument, which is a time series, and returns a **tuple** of six values:

- `rsquare` is a measurement for how well the line fits our original time series. The values of `rsquare` range from 0-1, where 0 denotes no fit and 1 indicates the best possible fit.

- `slope` is the slope of our best fit line.

- `variance` is the variance of the input data.

- `rvariance` is the variance between the original time series and the best fit line generated by `series_fit_line()`.

- `interception` is the interception point of the best fit line and the original time series.

- `line_fit` is the dynamic array of values we can pipe to `render` and plot as a chart.

The help cluster (`https://help.kusto.windows.net`) provided by Microsoft, which you can connect to via the ADX Web UI (`https://dataexplorer.azure.com`), contains a table called `demo_series3` that contains a time series with seasonality and with an upward trend.

This table contains the following three columns:

- `id` is an empty column.

- `t` contains a dynamic array of time values.

- `num` contains a dynamic array of the number of requests.

Since num is already a time series, we do not need to call make-series. We can simply call series_fit_line() and pipe the result to render timechart. The following query, which can be executed against the help cluster in https://dataexplorer. azure.com, plots the best fit line on our time chart:

```
demo_series3
| extend
(RSquare,Slope,Variance,RVariance,Interception,LineFit)=series_
fit_line(num)
| render timechart
```

The output of the preceding query is as follows:

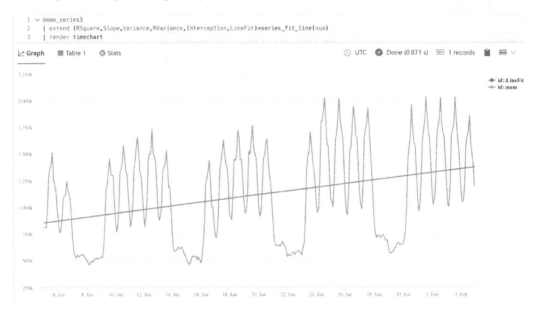

Figure 7.7 – Linear regression on demo_series3

As you can see, the straight line crossing the graph (id::LineFit), which is our best fit line, is showing an upward trend, which indicates that the number of requests is increasing. We can also look at the other values returned by series_fit_line() by projecting the values in tabular form.

The following query displays all the values returned by `series_fit_line()`:

```
demo_series3
| extend
(RSquare,Slope,Variance,RVariance,Interception,LineFit)=series_
fit_line(num)
| project RSquare,Slope,Variance,RVariance,Interception,LineFit
```

The value of `Rsquare`, which I mentioned earlier, is a standard measurement that ranges from 0 to 1, with 0 being the worst fit and 1 being the best fit; it represents the quality of the regression line. In our example, `Rsquare` is `0.12304991160566936`, which means the quality of the fit is relatively low and that future forecasting of the trend will not be highly accurate. The higher `Rsquare` is, the greater the accuracy of our estimated trends will be.

Like `series_fir()` and log analytics, we need to include two extra steps in our query to render the best fit line and original time series on a time chart. As you may recall from *Chapter 6*, *Introducing Time Series Analysis*, we looked at the number of security patches being installed. Let's extend this example and perform a linear regression to generate the best fit line.

The following query creates a time series for the last 100 days, plots the numbers of security updates that have been installed, and plots the best fit line. As we mentioned in the previous section, two extra steps are required to generate the graph in Log Analytics:

```
let startTime = ago(100d);
let endTime = now();
let binSize = 1d;
Update
| where Classification == "Security Updates"
| make-series updates=count() default=0 on TimeGenerated from
startTime to endTime step binSize
| extend
(RSquare,Slope,Variance,RVariance,Interception,LineFit)=series_
fit_line(updates)
| mv-expand LineFit to typeof(double), updates to
typeof(double), TimeGenerated to typeof(datetime)
| project TimeGenerated, updates, LineFit
| render timechart
```

The output of the preceding query is shown here:

```
1   let startTime = ago(100d);
2   let endTime = now();
3   let binSize = 1d;
4   Update
5   | where Classification == "Security Updates"
6   | make-series updates=count() default=0 on TimeGenerated from startTime to endTime step binSize
7   | extend (RSquare,Slope,Variance,RVariance,Interception,LineFit)=series_fit_line(updates)
8   | mv-expand LineFit to typeof(double), updates to typeof(double), TimeGenerated to typeof(datetime)
9   | project TimeGenerated, updates, LineFit
10  | render timechart
```

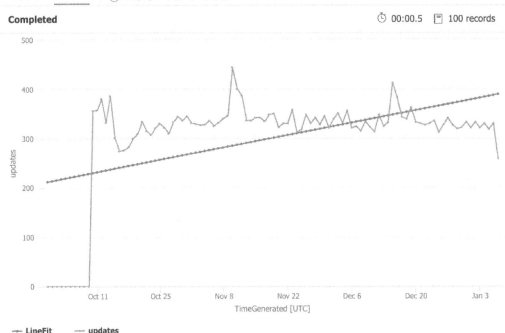

Figure 7.8 – Linear regression for the number of security patches installed

Here, *line 8*, | mv-expand LineFit to typeof(double), updates to typeof(double), TimeGenerated to typeof(datetime), uses mv-expand to split the array, while *line 9* projects the values we want to display in the time chart.

As shown in the preceding diagram, there is a slight upward trend. This tells us that there is, in general, an increase in the number of security updates being applied daily. We can then use this data in our planning to estimate how much time and resources we need to spend to apply the updates in our environment.

Applying segmented regression with KQL

There are times when your time series data may have a variance or step jump that makes it difficult to have a single best line of fit. **Segmented regression** is a method that can help you detect the optimum split point and create two best-fit lines – one to the left of the split point and another to the right of the split point.

KQL provides the `series_fit_2lines()` function to help you perform two-segment linear regression. The help cluster (`https://help.kusto.windows.net`) contains the `demo_series2` table to demonstrate two-segmented linear regression.

As shown in the following screenshot, there is a clear step jump in our data, which we will use to demonstrate `series_fit_lines()`:

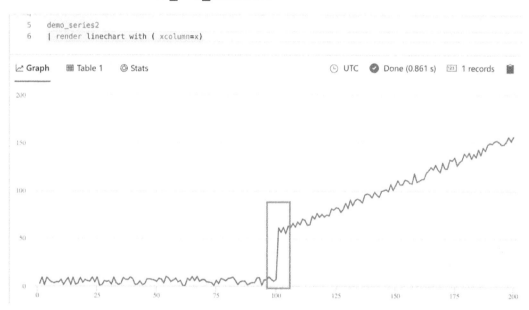

Figure 7.9 – A step jump in our time series

The `series_fit_lines()` function that we used earlier cannot detect such step jumps. Such step jumps are not uncommon and can degrade the quality of `rsquare`. `series_fit_2lines()`, on the other hand, analyzes the data, detects the split point, and creates two best lines of fit – one to the left of the split point and another to the right of the split point.

The following query calculates both the regular linear regression and the two-segment linear regression:

```
demo_series2
| extend series_fit_line(y), series_fit_2lines(y)
| render linechart with ( xcolumn=x, title="Two Segment Linear
Regression")
```

The results of this query are shown here:

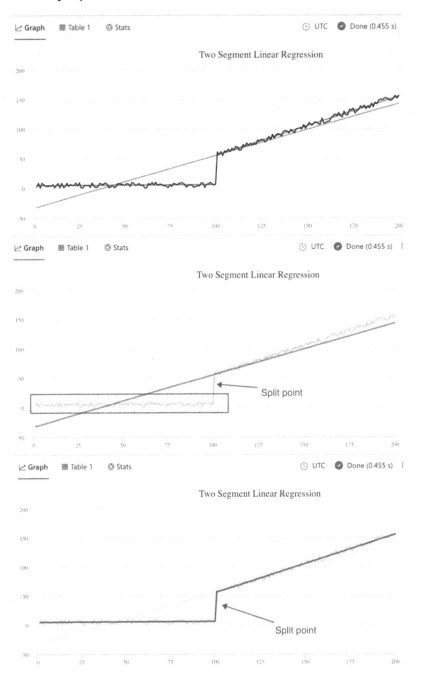

Figure 7.10 – Two-segment linear regression example

The preceding diagram illustrates three views of the data to make the best fit lines more readable:

- The first graph shows the original data, the single linear regression, and the two-segment linear regression.

- The second graph highlights the single linear regression. As you can see, the best fit line does not fit the time series on the left-hand side of the split point.

- The final graph highlights the two-segment best fit lines, which are near perfect.

In this section, we learned about the linear regression features of KQL and how we can use basic linear regression to identify trends and positive and negative relationships in our time series data. In the next section, *Anomaly detection and forecasting with KQL*, we will learn about KQL's advanced functions that allow us to perform anomaly detection and forecasting on our data.

Anomaly detection and forecasting with KQL

By now, you should have a good understanding of the different components of time series such as seasonality, trends, and variations. KQL provides the `series_decompose()` function to calculate the values of these components for a given time series.

The `series_decompose()` function expects one required argument and four optional arguments. Let's look at these arguments in more detail:

- `series` is the time series we would like to calculate the components for.

- `seasonality` is set to `-1` to have the function autodetect the seasonality, `0` to skip the seasonality analysis, or a positive integer to specify the expected period. The default value is `-1` (auto-detect).

- `trend` determines the type of trend analysis that's performed. There are three options we can specify at the time of writing:

  - `avg` specifies the average bins for the trend.

  - `linefit` specifies linear regression, which we learned about earlier, by using `series_fit_line()`.

  - `none` specifies that no trend analysis is performed.

- `test_points` specifies the number of values to exclude from the regression. When forecasting, we would like to decompose only the existing part of the signal and apply it for the future signals. In the anomaly detection, we may want to exclude the recent points that were tested in the decomposition process.

- `seasonality_threshold` specifies the threshold of the seasonality score. If the seasonality score is below this threshold, then the model assumes no seasonality.

- `series_decompose()` returns a four-value tuple:

  - `baseline` is the predicted value of the series.

  - `seasonal` is the seasonal component of the series.

  - `trend` is the trend component of the series.

  - `residual` is the result of the `baseline` value that's been subtracted from our time series.

Earlier in this chapter, we calculated the moving average for the number of HTTP requests from United States. Let's modify this query to render the time series components.

The following code queries the demo_make_series1 table from the help cluster to calculate and render the time series components. If you wish to execute the following query yourself, you can do so by executing the query on the help cluster via the ADX Web UI (https://dataexplorer.azure.com):

```
let startTime = toscalar(demo_make_series1 | summarize
min(TimeStamp));
let endTime = toscalar(demo_make_series1 | summarize
max(TimeStamp));
let binSize = 1h;
demo_make_series1
| where Country == "United States"
| make-series requests=count() default=0 on TimeStamp from
startTime to endTime step binSize by Country
| extend series_decompose(requests, -1, 'linefit')
| render timechart
```

The output of this query is shown here:

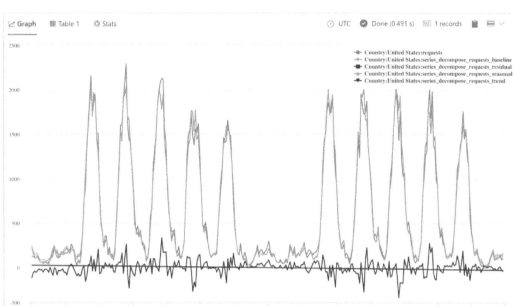

Figure 7.11 – Time series components rendered as a time chart

The preceding graph renders all the individual components of the time series returned by `series_decompose()`, which are the baseline, residual, seasonality, and the trend. In the next section, *Anomaly detection*, we will learn what anomaly detection is and how to apply KQL's anomaly functions.

Anomaly detection

The next function we will look at is `series_decompose_anomalies()`. The `series_decompose_anomalies()` function takes a time series as an argument and returns a three-value tuple that can be piped to the `render` operator.

Before diving into an example, let's understand what anomalies are. Per its definition, an anomaly is something such as a data point that is out of the norm. A good example of an event that was out of the norm and, in keeping with the football theme, was when Paris Saint-Germain bought Neymar from Barcelona for over two hundred million euros. This exorbitant amount was more than double the previous highest-paid total.

It is important to detect anomalies and reduce them by applying moving averages because anomalies can skew our forecasts. Remember, our goal is to ensure that the forecasts are as accurate as possible.

The following query performs anomaly detection on our HTTP request from the United States time series:

```
let startTime = toscalar(demo_make_series1 | summarize min(TimeStamp));
let endTime = toscalar(demo_make_series1 | summarize max(TimeStamp));
let binSize = 1h;
demo_make_series1
| where Country in ("United States", "Germany", "China")
| make-series requests=count() default=0 on TimeStamp from startTime to endTime step binSize by Country
| extend anomalies = series_decompose_anomalies(requests)
| render anomalychart with (anomalycolumns=anomalies)
```

The output of this query is shown here:

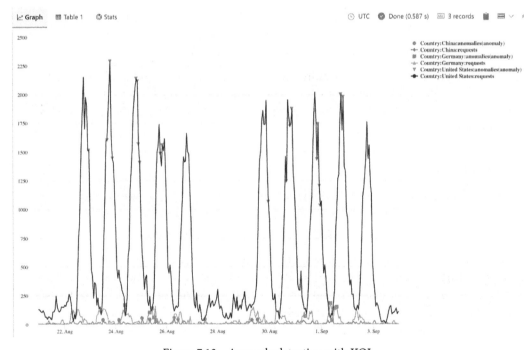

Figure 7.12 – Anomaly detection with KQL

As you can see, the anomalies are plotted as distinct dots on our graph. As we mentioned previously, the anomaly graph is not available in Log Analytics workspace at the time of writing.

Forecasting for the future

One of the goals of time series analysis is to be able to make forecasts and predictions for the future. The topics mentioned in the previous section help improve the accuracy of our forecasting by removing noise and anomalies from our data. KQL provides a function called `series_decompose_forecast()` that allows us to create forecasts on our data. The following query forecasts the number of HTTP requests for the next seven days:

```
let startTime = toscalar(demo_make_series1 | summarize
min(TimeStamp));

let endTime = toscalar(demo_make_series1 | summarize
max(TimeStamp));

let binSize = 1h;

let forecastSize = 7d;

demo_make_series1

| make-series requests=count() default=0 on TimeStamp from
startTime to endTime + forecastSize step binSize

| extend forecastedRequests = series_decompose_
forecast(requests, toint(forecastSize / binSize))

| render timechart
```

The output of this query is shown here:

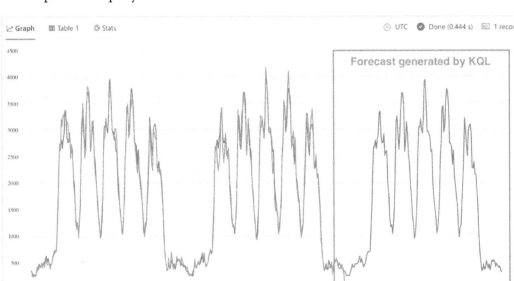

Figure 7.13 – Time series forecasting with KQL

As we can see, the forecast from September 4 to September 11 has been rendered on the time series graph and looks very similar to our actual data, before September 4. Now, let's break the query down line by line:

1. `let startTime = toscalar(demo_make_series1 | summarize min(TimeStamp));` returns the earliest date available in the dataset by using the `min()` aggregation function and assigns the result to the `startTime` variable. The `toscalar()` function returns a constant value of the nested summarized query.

2. `let endTime = toscalar(demo_make_series1 | summarize max(TimeStamp));` is similar to *line 1*, except it assigns the last date in the dataset and assigns the value to `endTime`.

3. `let binSize = 1h;` assigns the bin size to the `binSize` variable. In this case, we want a bin size of 1 day, which means we are capturing the total number of requests per day.

4. `let forecastSize = 7d;` is the length of the forecast. In this case, we want to generate a forecast for the next 7 days.

5. `demo_make_series1` is the data source we are querying.

6. `| make-series requests=count() default=0 on TimeStamp from startTime to endTime + forecastSize step binSize` generates our aggregated time series by counting the number of daily (`binSize`) requests.

7. `| extend forecastedRequests = series_decompose_forecast(requests, toint(forecastSize / binSize))` generates the 7-day forecast based on the existing data, excluding the empty future time points.

8. `| render timechart` renders the forecasted HTTP count as a time chart.

In this section, we learned how to perform anomaly detection and forecasting using KQL's decompose functions.

Summary

This chapter introduced the basics of time series analysis. For a deeper dive into time series analysis and statistics, I highly recommend looking at some of the great titles published by Packt, such as *Practical Time Series Analysis* and *Forecasting Time Series Data with Facebook Prophet*.

In this chapter, we learned about moving averages and how moving averages can help reduce noise and make our time series data smoother. Reducing noise helps us identify the patterns and common traits of time series data, such as variations and seasonality. Furthermore, reducing the noise helps improve our accuracy when making forecasts.

Next, we learned how to render moving averages and line regressions in Log Analytics. Log Analytics requires a couple of extra steps to be performed in the query before the data is rendered to the charts due to the Data Explorer Web UI and Log Analytics having different user agents. Please see `https://docs.microsoft.com/en-us/azure/data-explorer/kusto/query/renderoperator?pivots=azuredataexplorer` for more information.

Next, we learned how to perform trend analysis in KQL by using linear regression. Linear regression helps us understand the relationship between the independent variable (x-axis) and the dependent variable (y-axis) and identify positive and negative trends.

Finally, we learned how we can determine the seasonality, trend, and residual components of our time series using `series_decompose()`, as well as how to perform anomaly detection using `series_decompose_anomalies()` and forecasting using `series_decompose_forecast()`. It's worth mentioning that these functions can be applied and calculated for over thousands of time series in seconds. In this chapter, our examples only used a single time series for simplicity and readability.

In the next chapter, *Data Visualization with ADX and Power BI*, we will learn how to create reports and dashboards.

Questions

Before moving on to the next chapter, test your knowledge by answering these questions. The answers can be found at the back of this book:

1. What is the purpose of moving averages?

2. What is the purpose of linear regression?

3. What are the extra steps required to render time charts in log analytics?

4. In *Figure 7.12*, we rendered an anomaly chart to display the anomalies in the time series. Using `series_fir()`, generate a smoother graph without the anomalies. Once you have generated a smoother output, pass your data to `series_decompose_anomalies()` to see if there are still any anomalies. The query for generating the graph in *Figure 7.12* is as follows. You will need to connect to the help cluster (`https://help.kusto.windows.net/`) to complete this exercise:

```
let startTime = toscalar(demo_make_series1 | summarize
min(TimeStamp));

let endTime = toscalar(demo_make_series1 | summarize
max(TimeStamp));

let binSize = 1h;

demo_make_series1

| make-series requests=count() default=0 on TimeStamp
from startTime to endTime step binSize

| extend anomalies = series_decompose_anomalies(requests)

| render anomalychart with (anomalycolumns=anomalies)
```

8
Data Visualization with Azure Data Explorer and Power BI

By this point, you should have a solid understanding of **Azure Data Explorer** (ADX), understand how to deploy ADX infrastructure, ingest data, how to query your data using KQL, and how to perform time series analysis and create forecasts. Once you understand your data, the next step is to present the findings to your key stakeholders. We can apply the concepts of **data visualization** using ADX and **Power BI**.

In this chapter, we will begin by introducing the concept of data visualization. We will discuss some of the important questions and design principles that should be taken into consideration to ensure the narrative is clear, concise, and data-driven rather than being based on bias and gut feeling. We will also discuss some of the different chart types and when to use them.

Next, we will introduce the dashboard capabilities of the **Azure Data Explorer Web UI** and learn how to navigate the **dashboard** window, how to create dashboards, and how to apply the design principles discussed previously. Armed with this knowledge, we will understand how to share dashboards and how to make dashboards interactive by creating filters that can be used by your stakeholders.

The rest of this chapter will focus on integrating ADX with **Power BI** and explain how to create a work account using **Azure Active Directory**, how to connect our ADX instance to Power BI as a data source, and how to create reports using our ADX data as our source. I will demonstrate this by creating a report in Power BI to render the storm locations on a map. With this map, I will display the different types of storm events for a given state in the United States of America.

In this chapter, we are going to cover the following main topics:

- Introducing data visualization
- Creating dashboards with Azure Data Explorer
- Connecting Power BI to Azure Data Explorer

Technical requirements

The code examples for this chapter can be found in the `Chapter08` folder of this book's GitHub repository: `https://github.com/PacktPublishing/Scalable-Data-Analytics-with-Azure-Data-Explorer.git`.

In our examples, we will use the `StormEvents` table in our ADX cluster, since this dataset provides a wide variety of data types that allow us to demonstrate multiple types of visualizations.

You will need a Power BI account. Power BI requires a work or school account, which means that email addresses such as outlook.com and gmail.com will not work. Azure Active Directory accounts are work accounts and, in this chapter, we will discover how to create Azure Active Directory accounts to sign up to Power BI. You will also need Windows since the examples in this chapter require the Power BI desktop application, which is currently a Windows-only application.

Introducing data visualization

Before diving into building dashboards, it is worth spending some time discussing what data visualization is and its goals. As we mentioned in *Chapter 1*, *Introducing Azure Data Explorer*, 90% of today's data is digital and we are generating quintillion bytes of data each day!

Once we have understood our data and identified traits such as trends, variations, seasonality, and anomalies and created forecasts with them, the next step is to present our findings to our audience. This is where data visualization can help. Data visualization is a method that helps facilitate your understanding of your data to your audience, who can have various backgrounds and expertise.

Designing and developing effective data visualizations is an art and requires practice. The types of charts and tiles you use can influence how your data is perceived and bad design decisions could lead to your audience interpreting the data incorrectly.

To illustrate the power of data visualization, let's assume we want to show the number of injuries, deaths, cost of property damage, number of events, and locations of events for California. The following query in ADX returns all the requested information represented in a table:

```
StormEvents
| where State == "CALIFORNIA"
| project EventType, InjuriesDirect, DeathsDirect,
DamageProperty, BeginLat, BeginLon
```

All the data that was requested is present, as shown in the following screenshot, but it is difficult to comprehend. So, without having any prior understanding of the data, it is difficult to draw conclusions. For instance, consider the BeginLat and BeginLon columns. Unless you are an expert, inferring the states from BeginLat and BeginLon is next to impossible:

EventType	InjuriesDirect	DeathsDirect	DamageProperty	BeginLat	BeginLon
Flash Flood	0	0	0	33.334	-116.9205
Thunderstorm Wind	0	0	15,000	32.5709	-117.0605
Heavy Rain	0	0	0	38.52	-121.5
Tornado	0	0	2,000	38.3911	-121.37
Heavy Rain	0	0	0	40.5	-122.3
Flood	0	0	0	35.0482	-119.3801
Hail	0	0	0	34.13	-116.05
Flash Flood	0	0	0	34.1411	-116.0177
Hail	0	0	0	36.2344	-119.4141
Hail	0	0	0	36.2446	-119.57
Thunderstorm Wind	0	0	6,000	39.92	-122.1234
Hail	0	0	0	38.22	-119.02
Flash Flood	0	0	0	34.58	-118.1024
Funnel Cloud	0	0	0	32.8602	-117.2822
Heavy Rain	0	0	0	33.62	-114.6

Figure 8.1 – Storm events for California

On the other hand, if we apply our data visualization tools and principles, we can create dashboards that are a lot easier to comprehend, without requiring the audience to be familiar with the data, as shown in the following screenshot:

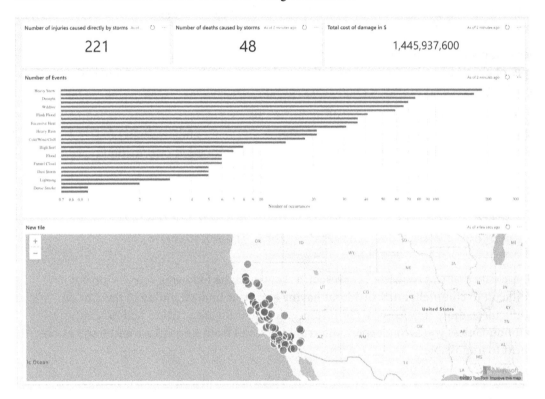

Figure 8.2 – Data visualization dashboards

As you can see, by leveraging some of the different chart types, we can make the data jump out and speak for itself.

In the next section, *Creating dashboards with Azure Data Explorer*, we will learn how to develop and share dashboards in the Data Explorer Web UI.

Creating dashboards with Azure Data Explorer

In this section, we will learn how to navigate the dashboard editor, how to build basic dashboards with various charts and tiles, how to share dashboards, and how to develop dashboards with parameters that allow your audience to interact with and change parameters. As you will see later in this chapter, parameters allow us to create predefined filters that the audience can use to manipulate their view of the dashboard.

Navigating the dashboard window

At the time of writing, the Data Explorer Web UI's navigation panel contains three options called **Data**, **Query**, and **Dashboards (Preview)**, as shown in the following screenshot:

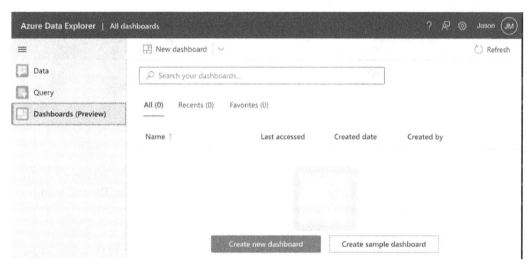

Figure 8.3 – The ADX Web UI's navigation panel

The first time you click on **Dashboards (Preview)** in ADX (`https://dataexplorer.azure.com`), you will see two buttons: **Create new dashboard** and **Create sample dashboard**. The **Create sample dashboard** button creates a dashboard with some preconfigured tiles to demonstrate the different capabilities of dashboards.

To create our first dashboard, click **Create new dashboard**. As shown in the following screenshot, you will be prompted to give your dashboard a name. Enter `Storm Summary for California` and click **Create**:

Figure 8.4 – Creating a new dashboard

When you click **Create**, you will be taken to the dashboard editor window. Before we build a dashboard, let's review the different controls and buttons in editor mode. The following screenshot shows each component of the editor window:

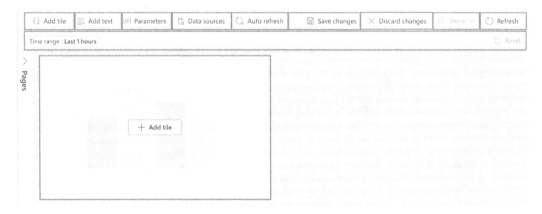

Figure 8.5 – Dashboard editor mode

The top menu bar consists of nine options:

- The **Add tile** option allows us to add additional tiles to the dashboard. As we will soon see, tiles use KQL queries.

- The **Add text** option allows us to add a tile that contains text. This text can be written in Markdown, which is a popular language that supports various formatting options.

- The **Parameters** option, as we will see later in this chapter, allows us to define filters that make our dashboards interactive.

- The **Data Sources** option is where we specify the ADX cluster and database.

- The **Auto refresh** option allows us to configure the auto-refresh options for the dashboard. If you have dynamic data, it is good to auto-refresh to ensure you are displaying the latest information.

- The **Save changes** option allows us to save any changes we have made to the dashboard.

- The **Discard changes** option allows us to discard any changes we have made to the dashboard.

- The **Share** option allows us to share the dashboard with other users.

- The **Refresh** option allows us to refresh the dashboard and pull the latest data.

The second menu contains the list of filters/parameters we have defined. We will define a state parameter later in this chapter to demonstrate the use of filters. By default, there is a time range filter. The **reset** button reverts any filters that you apply to the dashboard.

The next portion of the window is our dashboard canvas and contains our tiles. We can add as many tiles as we like and can arrange them in any order.

In the next section, *Building our first Data Explorer dashboard*, we will create the dashboard that was presented earlier in this chapter (*Figure 8.2*), which showed a summary of storm events for California.

Building our first Data Explorer dashboard

The following steps describe how to create a Data Explorer dashboard that consists of a text tile, three stats tiles, a bar chart, and a map:

1. The first task is to connect the dashboard to our ADX cluster and database. Click **Data sources**.

2. Enter StormEvents for **Data source name**.

3. Enter your ADX cluster's URL in the **Cluster URI** field – for example,
    `https://adxmyerscough.westeurope.kusto.windows.net/` –
    and click **Connect**.

4. From the drop-down list, select the name of your database – for example, `adx-db` –
    and click **Apply**, as shown in the following screenshot:

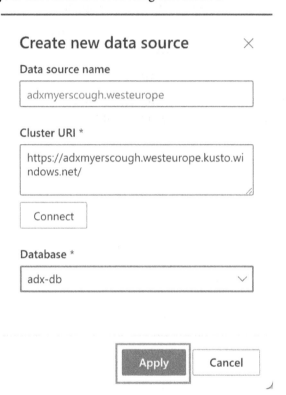

Figure 8.6 – Adding a data source

5. Let's begin by adding a text tile that will contain the title and a summary of our
    dashboard. Click **Add text**.

6. Enter the following Markdown text and click **Apply changes**:

```
# Storm Summary for California
_The following dashboard provides a summary of the
direct injuries and deaths, total property damage, an
aggregation of all the storm events in California and the
location of each storm._
```

7. Resize the new tile so that it spans the top of the dashboard, as shown in the following screenshot:

Figure 8.7 – Creating the title tile

8. The next step is to add a tile to display the direct injuries. Click **Add tile**. A query editor will appear. As I mentioned earlier, the tiles are powered by KQL queries. Enter the following query to aggregate the total number of direct injuries:

```
StormEvents
| where State == "CALIFORNIA"
| summarize DirectInjuries=sum(InjuriesDirect)
| render card with (xtitle="Number of injuries caused
directly by storms")
```

9. Next, run the query by clicking **Run**. Then, click **+ Add visual** to open the **Visual formatting** panel.

10. Set **Visual type** to **Stat**. Then, under **General**, set **Text size** to **Large**, as shown in the following screenshot. The **Stat** visual type allows us to display a single value, such as a counter value:

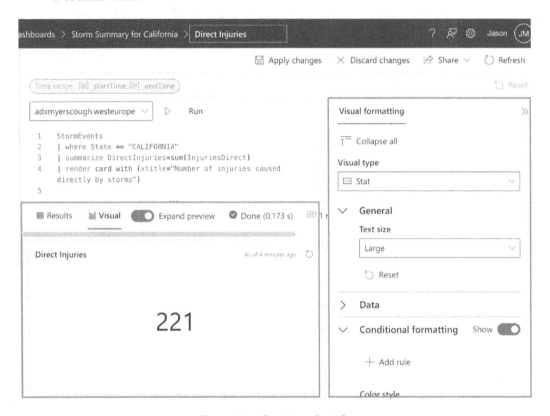

Figure 8.8 – Creating a Stat tile

11. All tiles have a title, as shown at the top of the preceding screenshot, and there is also a crumb trail and a textbox. Enter `Direct Injuries` as the name and click **Apply changes**.

12. Resize the tile and position it under our title tile, as shown in the following screenshot:

Figure 8.9 – Direct Injuries tile

13. Next, let's add a **Stat** tile to display the total number of direct deaths.

14. Click **Add tile** and enter the following query:

```
StormEvents
| where State == "CALIFORNIA"
| summarize DirectInjuries=sum(DeathsDirect)
```

15. Run the query by clicking **Run** and then click **+ Add visual**.

16. Set **Visual style** to **Stat**. Then, under **General**, set **Text size** to **Large**.

17. Name the tile **Direct Deaths** and click **Apply changes**.

18. Now, let's add the final **State** tile to display the total cost of damage. Click **Add tile** and enter the following query:

```
StormEvents
| where State == "CALIFORNIA"
| summarize TotalDamage=sum(DamageProperty)
```

19. Run the query and click **+ Add visual**.

20. Set **Visual type** to **Stat**. Then, under **General**, set **Text size** to **Large**.

21. Enter **Total Cost** as the title and click **Apply changes**.

So far, we have added three **Stat** tiles to our dashboard. The following steps explain how to add the bar chart for event types and a map to display the location of storms:

1. Click **Add tile** and enter the following query:

```
StormEvents
| where State == "CALIFORNIA"
| summarize _count=count() by EventType
| order by _count desc
```

2. Run the query by clicking **Run** and click + **Add visual**.

3. Set **Visual type** to **Bar chart** and turn off the legend.

4. Name the tile **Event Type Aggregation** and click **Apply changes**.

5. Let's add our final tile – the map tile – by clicking **Add tile** and entering the following query:

```
StormEvents
| where State == "CALIFORNIA"
| where isnotempty( BeginLat) and isnotempty( BeginLon)
```

6. Run the query and click + **Add visual**.

7. Set **Visual type** to **Map**, name the tile **Event Locations**, and click **Apply Changes**.

8. Under the **Data** category, set the value of **Define location by**: to **Latitude and longitude**.

9. Set **Latitude column** to BeginLat (double) and set **Longitude column** to **BeginLon (double)**.

10. Finally, click **Save changes** to save the dashboard. Your dashboard should look similar to the following:

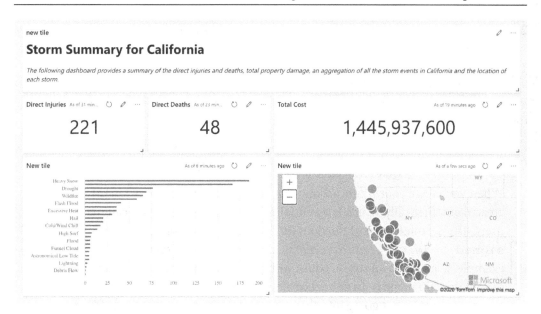

Figure 8.10 – Our first Data Explorer Dashboard

> **Note**
>
> When developing dashboards, please remember and get into the habit of saving your changes regularly to avoid losing changes.

In the next section, *Sharing dashboards*, we will learn how to share dashboards with other people.

Sharing dashboards

By default, dashboards are not visible to other users. To share our dashboards, we must explicitly grant permission at the dashboard level and the ADX cluster level. We will discuss user permissions in *Chapter 10*, *Azure Data Explorer Security*. For now, do not worry about the details. In this section, we will create an Azure Active Directory user that we can use to share our dashboards and use Power BI. Let's get started:

1. Navigate to the Azure portal (`https://portal.azure.com`).

2. Click **All services**, search for Azure Active Directory, and click **Azure Active Directory**, as shown here:

Figure 8.11 – Searching for Azure Active Directory

When you create your Azure account, an **Azure Active Directory tenant** will be created automatically, which is essentially your domain where you can manage users. The default format of the domain name is <your_email_address_you_signed_up_with>.onmicrosoft.com, minus the @ symbol and periods. For example, I created my Azure account with jason.adx@outlook.com and my default domain name is jasonadxoutlook.onmicrosoft.com. You can use a custom domain, but going into this is beyond the scope of this book. Please see https://docs.microsoft.com/en-us/azure/active-directory/fundamentals/add-custom-domain if you are interested in creating a custom domain.

When you create a new user, their username, also known as the user service principal, is in the <name>@<primary domain> format; for example, james@jasonadxoutlook.com.

3. In the **Properties** pane, under **Manage**, click **Users** and click **+New User**.

4. Enter the name of the user you want to create. For example, I have created a user called james, as shown in the following screenshot:

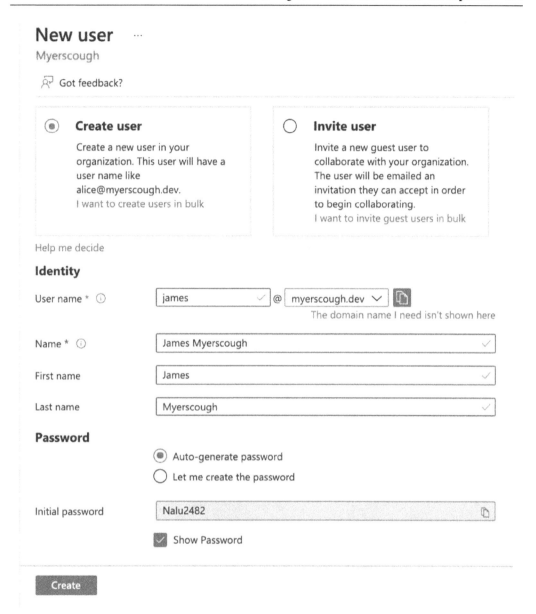

Figure 8.12 – Creating a new Azure AD user

You may have noticed that my domain name looks different from what I described earlier. This is because I have created a custom domain.

5. Ensure you tick **Show Password**. This is a one-time password, and you will be asked to change it when you log in for the first time.

6. Click **Create** to create the user account.

As we mentioned earlier, user management will be discussed in more detail in *Chapter 10, Azure Data Explorer Security*.

The following steps demonstrate how to assign an Azure Active Directory user read permission to our ADX cluster and how to share our dashboard:

1. Open the Azure portal (`https://portal.azure.com`).

2. Click on your ADX cluster – for instance, `adxmyerscough`. Then, in the **Properties** panel, click **Permissions**, then **+ Add**, and then **AllDatabasesViewer**. Search for a user in your AD tenant.

3. Click the user you want to share your dashboard with and click **Select**, as shown in the following screenshot:

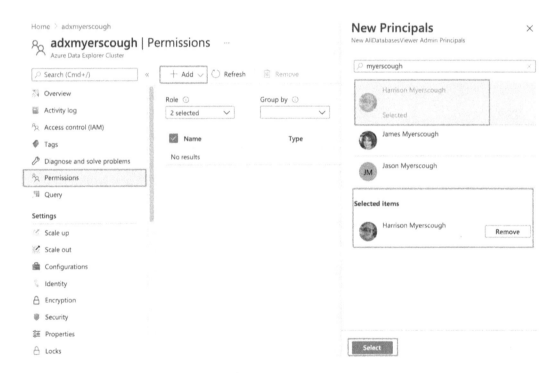

Figure 8.13 – Adding users to the ADX cluster

4. Go back to the Data Explorer Web UI and click **Share** and then **Manage Permissions**.

5. Now, enter the username or email of the person you would like to share the dashboard with. For instance, I am going to share my dashboard with `harrison@myerscough.dev`.

6. You have the option to give read or edit permissions. For now, keep the permission set to **Can view** and click **Add**.

7. Click **Copy link** to copy the link to your clipboard. You can now share this link with your user. In my case, the URL is `https://dataexplorer.azure.com/dashboards/9b7a5bdb-30c2-448c-a61c-5809a3881608`.

As shown in the following screenshot, Harrison can view the dashboard but is unable to make any edits:

Figure 8.14 – Harrison has read-only permissions

If you forget to add the user in the AD tenant, the user will see the following errors when you share the link:

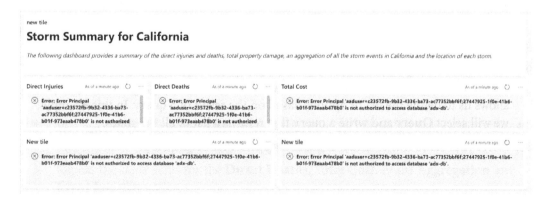

Figure 8.15 – Insufficient permissions

4. Now, select a different state. You should see your tiles update according to which state you have selected. For example, the following screenshot shows a summary of ILLINOIS:

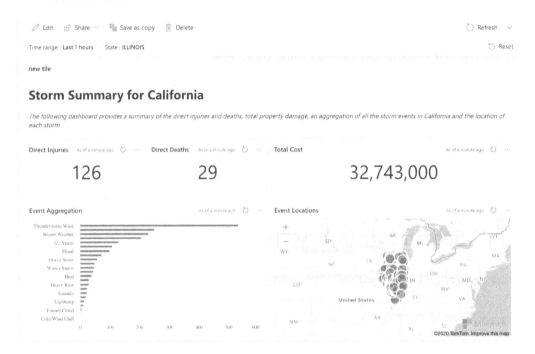

Figure 8.17 – Dashboards with filters

If you click **Reset**, the filter will be reverted to the default value, which in our case is CALIFORNIA.

You may have noticed that the title still says California, regardless of which state you have selected. This is because markup text fields are static. For completeness, let's update the title. The following steps demonstrate how to update the text field:

1. Click **Edit** and then click the text field's pencil icon at the far right to enter edit mode:

Figure 8.18 – Editing the text field

2. Enter the following Markdown:

```
# Storm Summary for the United States
_The following dashboard provides a summary of the
direct injuries and deaths, total property damage, an
aggregation of all the storm events in the United States
and the location of each storm._
```

3. Click **Apply changes** and then click **Save changes** to save the dashboard.

As shown in the following screenshot, our dashboard looks more professional:

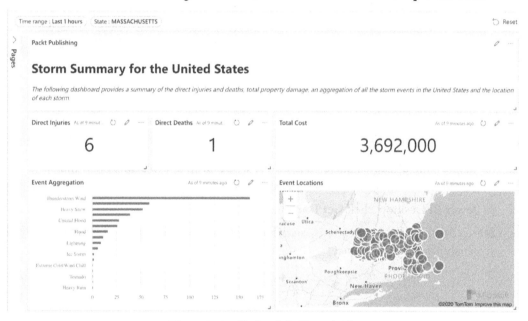

Figure 8.19 – Updated dashboard title

In this section, we reviewed how to create a dashboard in the Data Explorer Web UI with and without parameters. We also learned how to share these dashboards with other users. In the next section, *Connecting Power BI to Azure Data Explorer*, we will learn how to integrate ADX with Power BI.

Connecting Power BI to Azure Data Explorer

Power BI is Microsoft's enterprise business intelligence product. It allows us to build powerful reports that give deep insights into our data that we can use to make data-driven decisions.

This section is not intended to be an introduction to Power BI; I assume you are already familiar with Power BI and would like to integrate ADX with Power BI. From my experience, most non-technical personnel will not log into portals such as the Azure portal or the Data Explorer Web UI. Instead, they will prefer to use tools they are familiar with and there is a high chance they are using Power BI. This section will explain how to create reports based on your ADX data and share it with a wider audience, such as product management teams.

With that said, it is only possible to sign up to Power BI using either a school or work email address. Email addresses such as gmail.com, icloud.com, and so on are not valid. The good news is that your Azure Active Directory users are considered work emails and it is possible to sign up to Power BI using one of those. For this demonstration, I signed up to Power BI using the `harrison@myerscough.dev` account. Please go ahead and sign up to `https://powerbi.microsoft.com/` using the Azure AD account you created earlier.

The following example demonstrates how to connect ADX instances to your Power BI workspaces. Unfortunately, at the time of writing, the ADX connector is only available from within the Power BI Desktop application, which is a Windows-only application:

1. If you have not already installed the **Power BI desktop application**, open the **Windows Store**, search for Power BI, and install it.

2. In Power BI Desktop, click **Get Data** and then **More…**.

3. Search for the ADX connector by searching for `kusto`, as shown in the following screenshot, and click **Connect**:

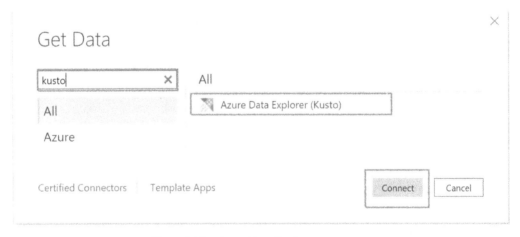

Figure 8.20 – Azure Data Explorer connector

4. Next, add the URL for your cluster; for instance, `https://adxmyerscough.westeurope.kusto.windows.net`.

5. Keep **Data Connectivity mode** set to **Import** as this will import our data into Power BI. Since our data does not change frequently, importing is sufficient for our needs. If you have a dataset that is too large to be imported into Power BI or have a dataset that changes frequently, then **DirectQuery** is the preferred connectivity mode.

6. Click **OK**.

7. Next, you will be prompted to sign into Azure. Enter your credentials and click **Connect**.

8. Select the `StormEvents` table and click **Load**. This will import the entire dataset into Power BI, as shown in the following screenshot:

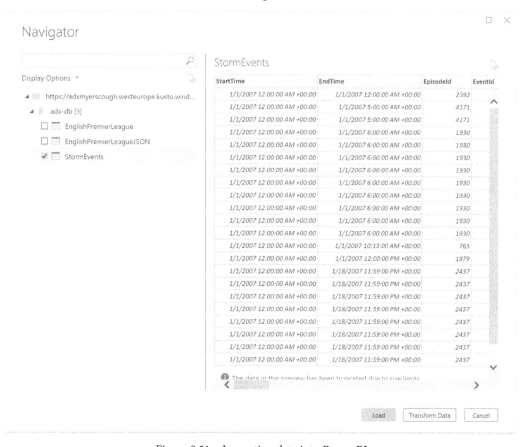

Figure 8.21 – Importing data into Power BI

The import can take up to a couple of minutes, depending on your internet speed and the response time from Azure. Once the table has been imported, you can create your Power BI reports using the data from your ADX cluster.

Next, let's create a simple Power BI report containing a map to display the storm locations and a pie chart to display the types of events.

9. Under **Visualizations**, click **Map** to place a map tile on your report.

10. Under **Fields**, expand the StormEvents table and drag the BeginLat column over to the Latitude property, below **Visualizations**.

11. Drag the BeginLon column over to the Longitude property, as shown in the following screenshot:

Figure 8.22 – Creating a report in Power BI

12. Click your report canvas to shift the focus to the report and click **Pie Chart** from the **Visualizations** options.

13. Now, drag the EventType column over to the Details and Values properties, as shown in the following screenshot:

Figure 8.23 – Pie chart properties

The following screenshot shows the basic report we just created in Power BI using the data from our ADX cluster:

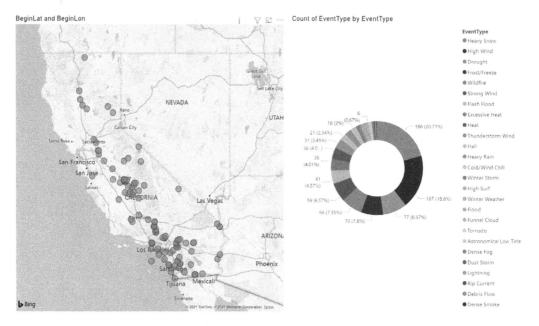

Figure 8.24 – Generating reports in Power BI using data from ADX

As you can see, we were able to take the raw longitude and latitude values and render them on a map. This not only helps display the locations but there is also no prerequisite for your audience to understand longitude and latitude values.

In this section, you learned how to connect Power BI to ADX and discovered how to create basic reports using some of the widgets available in Power BI. Remember, the types of widgets you use to display your data depends on the type of data and the message you want to convey. You are now equipped to import any of your ADX datasets into Power BI and share those with your audience.

Summary

In this chapter, we began by introducing the concept of data visualization and discussed some of the important questions and design principles that should be taken into consideration to ensure the narrative is clear, concise, and data-driven rather than being based on bias and gut feeling. We discussed some of the different chart types and when to use them.

Next, we introduced the dashboard capabilities of the Azure Data Explorer Web UI and learned how to navigate the dashboard window, how to create dashboards and apply the design principles discussed in the previous section, and then understood how to share dashboards and how to make dashboards interactive by creating filters, which can be used by your stakeholders.

Finally, we explained how to integrate ADX with Power BI and how to create a work account using Azure Active Directory, as well as how to connect our ADX instance to Power BI as a data source. We created a report in Power BI to display the storm event locations on a map. This was a good example of simplifying the interpretation for the audience. For instance, viewing points on a map is a lot easier than trying to understand the raw data of longitude and latitude coordinates.

In the next chapter, *Monitoring and Troubleshooting Azure Data Explorer*, we will investigate how we can monitor our ADX instances and troubleshoot issues such as data ingestion problems.

Questions

Before moving on to the next chapter, test your knowledge by answering the following questions. The answers can be found at the back of this book:

1. What is data visualization?
2. Create a dashboard in Data Explorer for our `EnglishPremierLeague` table and display the total number of goals scored, the goals conceded, and the number of wins for each team. Experiment and add some extra tiles.
3. What is the purpose of parameters in Data Explorer dashboards?
4. What is the difference between **Import** and **DirectQuery** when configuring the ADX connector for Power BI?
5. Try modifying the dashboard we created in the Data Explorer Web UI and update the title by editing the Markdown text.

Section 3: Advanced Azure Data Explorer Topics

This section of the book focuses on some of the advanced **Azure Data Explorer** (ADX) topics, starting with monitoring and troubleshooting your ADX clusters. It then discusses how to secure your ADX environment by restricting access and applying the principle of least privilege using **role-based access control** (**RBAC**) and deploying your clusters in virtual networks. The final two chapters discuss some of the best practices for improving your KQL query performance and how to manage your cluster performance and monthly ADX costs by explaining how to size your ADX clusters correctly, how to manage your cluster scaling, and data retention periods.

This section consists of the following chapters:

- *Chapter 9, Monitoring and Troubleshooting Azure Data Explorer*
- *Chapter 10, Azure Data Explorer Security*
- *Chapter 11, Performance Tuning in Azure Data Explorer*
- *Chapter 12, Cost Management in Azure Data Explorer*

9
Monitoring and Troubleshooting Azure Data Explorer

Monitoring systems and environments seems to be more of an afterthought rather than a part of initial requirements and design. During postmortems for production issues, it is not uncommon to have a long list of action items to remediate basic monitoring that we might assume is already implemented, such as **Secure Sockets Layer** (**SSL**) certificate expiration. Non-production environments are another story; it is not uncommon to find environments such as **user acceptance testing** (**UAT**), **quality assurance** (**QA**), and staging environments with no monitoring. Avoid these bad practices; always monitor your resources, regardless of the environment. How you raise an alert can differ depending on the environment and your service-level agreements (SLAs), but ensure that you always monitor your resources; otherwise, you are setting yourself up for failure.

In this chapter, we will begin by introducing the concepts of monitoring in **Azure Data Explorer** (**ADX**), discussing its importance, and understanding the significance of **service-level indicators** (**SLIs**), **service-level objectives** (**SLOs**), and **service-level agreements** (**SLAs**). Then, we will introduce the concept of troubleshooting and discuss my thought process and how we can break down problems.

The rest of the chapter will focus on the key metrics and logs available to us for ADX, and we will work through an example and troubleshoot an issue where data ingestion is not working. Finally, we will configure an alert so that we are notified of ingestion failures.

In this chapter, we are going to cover the following main topics:

- Introducing monitoring and troubleshooting
- Monitoring ADX
- Troubleshooting ADX

Technical requirements

The code examples for this chapter can be found in the Chapter09 folder of our repository at the following link: https://github.com/PacktPublishing/Scalable-Data-Analytics-with-Azure-Data-Explorer.git.

In our examples, we will be using the EnglishPremierLeagueJSON table we created in *Chapter 4*, *Ingesting Data in Azure Data Explorer*, and the storage account, event grid, and event hub that we deployed. If you have deleted those resources, please redeploy your infrastructure before continuing.

Introducing monitoring and troubleshooting

Before diving into monitoring and troubleshooting ADX, it is worth spending some time introducing the concepts of monitoring and troubleshooting.

There is no consensus on one definition for monitoring. Engineers have various backgrounds and different interests, and if you were to ask 10 engineers for a definition of monitoring, you would probably get 15 different answers. From my perspective, monitoring is a tool that aids with troubleshooting and allows us to measure and observe system behavior. I like to use the analogy of a compass, whereby monitoring is leading us to issues and giving insights into overall behavior, health, and performance.

Monitoring can typically be broken down into four functions: alerting, debugging, trends, and plumbing. I tend to agree with this but would like to extend on the plumbing aspects. Here's an overview of these functions:

- **Alerting**: Being able to notify engineers when issues occur. There is also an option to invoke self-healing capabilities, which Azure's alerting framework supports.

- **Debugging/troubleshooting**: The ability to access performance metrics and logging data to help support you in your troubleshooting activities.

- **Trends**: Collections of time series data such as **central processing unit (CPU)**, **random-access memory (RAM)**, and disk consumption, and using this data to identify trends such as growth, which you can then use in your decision-making process for areas such as capacity management.

- **Plumbing**: Brian Brazil, the creator of **Prometheus**, gives an overview of data pipelines and so on for ingestion and collecting the telemetry in his book *Prometheus Up & Running*. I would like to extend this and talk about three important components that should be considered when defining your monitoring: SLIs, SLOs, and SLAs.

In a nutshell, SLIs, SLOs, and SLAs are the requirements for your monitoring needs. They define the metrics you are interested in (SLIs), the thresholds for your alerting criteria (SLOs), and the service you are providing to your customer (SLAs)—for example, an SLA could be ensuring your service is available 99.99% of the time. Let's look at these in more detail here:

- **SLIs**: SLIs are the individual indicators that you are monitoring. An SLO could depend on multiple SLIs. In the context of ADX, a few examples of SLIs are CPU, ingestion latency, and hot cache size. We will cover more ADX metrics in the next section, *Monitoring ADX*.

- **SLOs**: SLOs are typically internal goals or objectives for a team—for example, ensuring the response time is less than 5 **milliseconds (ms)**. SLOs are essentially the requirements for your alerting thresholds and conditions.

- **SLAs**: SLAs are the agreements you make with your customers and are typically put into contracts. For example, ensuring the uptime of your system is a common SLA.

One of the keys to having good monitoring is to define your SLIs and SLOs early in your analysis and design phases.

Another important topic is troubleshooting. Troubleshooting is one of those topics that are difficult to teach. A lot of literature and learning resources tend to focus on specific tools; while this is helpful, it requires you to understand what type of issue you are dealing with. In my opinion, it is more important to learn a troubleshooting process based on problem solving and the process of elimination by quickly ruling out variables and factors that could be causing a problem. When I encounter a problem, I use the following process to help eliminate and home in on an issue:

1. **Understanding the problem**: This may sound obvious, but it is important to understand what the issue is and what the current impact is on customers. This helps identify a starting point for your investigation.

2. **Process of elimination**: Large, complex systems can be overwhelming. The key to problem solving is to break the problem down into smaller chunks and rule out possible factors/variables. Monitoring is a key component when it comes to the process of elimination. With our monitoring and telemetry, we can make data-driven decisions that allow us to make decisions quickly.

In the final section in this chapter, *Troubleshooting ADX*, we will work through an example of using this process of elimination to solve a problem with data ingestion.

In this section, I introduced the concepts of monitoring and troubleshooting and why they are important. In the next section, *Monitoring ADX*, we will learn which metrics and logging are available to us and how we can monitor our ADX instances using insights, log analytics, dashboards, and workbooks.

Monitoring ADX

During my time using the Azure platform, one of the areas I have seen make a lot of changes is monitoring and security. Almost every resource type has monitoring and security options in their properties panel. In *Chapter 10*, *Azure Data Explorer Security*, we will look at the security options, but for now, let's focus on the monitoring aspects.

Azure Service Health

Before jumping into metrics and logging, I think it is worth mentioning the **Service Health** blade. From my experience, not a lot of people are aware of the **Service Health** blade and what it offers. The **Service Health** blade offers a high-level overview of historical issues, current issues, planned maintenance, and security advisories. Another nice feature is that **Microsoft** posts its **root cause analysis (RCA)** reports here. The RCA reports provide a detailed description and timeline for issues, along with mitigation steps and follow-up action. The **Service Health** blade is one of the first things I check whenever an issue is brought to my attention.

The **Service Health** blade can be found by doing the following:

1. Logging in to the Azure portal (`https://portal.azure.com`.
2. Clicking **All Services** and searching for `Service Health`

The Azure **Service Health** blade is shown in the following screenshot:

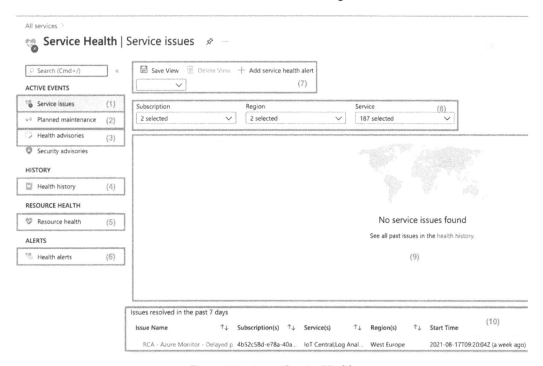

Figure 9.1 – Azure Service Health

Let's review the different sections of the **Service Health** blade, as follows:

1. **Service issues** displays a list of known issues currently impacting Azure. As you can see in *Figure 9.1*, there are no issues at the time of writing.
2. **Planned maintenance** displays a list of scheduled maintenance activities Microsoft has planned—for example, updating its Microsoft SQL version for its Azure SQL service.
3. **Health advisories** shows changes to Azure services that you are using—for example, if you have resources deployed in the legacy **Classic Mode**, you will receive notifications that the resource should be upgraded.
4. **Health history** displays a list of historical issues for the last 3 months.

5. **Resource health** provides an overview of your deployed Azure resources, such as ADX. The aforementioned views display platform-wide issues, but resource health is unique to your deployments.

6. **Health alerts** displays a list of health-related alerts that you have configured.

7. The **Context** menu allows you to save your views—for example, you can use the filters and select just the Azure regions and resource types that you are interested in. This is useful, as otherwise, you would have to configure the view every time you open the **Service Health** blade. You can also delete your saved views and create health-related alerts, such as alerting when there is a problem with a particular resource source type or when there is planned maintenance.

8. Filters, as mentioned in the preceding point, allow us to customize our views. For example, we can refine the number of Azure regions we are interested in and the types of Azure resources, such as **Virtual Machines** (**VM**) or ADX.

9. List of current issues displays a list of issues, if there are any.

10. RCA reports are useful, especially when you are impacted by an issue. These reports give a detailed report of what happened and what will be done to ensure the issue does not happen again.

Whenever you experience issues with your Azure resources, I highly recommend you use the **Service Health** blade as your first point of reference. I also recommend that you configure alerts for your critical resources, as we will learn later in the chapter.

In the next section, *ADX metrics*, we are going to discover the metrics that are available to us out of the box and how we can plot data on graphs.

ADX metrics

Metrics are our SLIs/raw data that we can use to measure the usage and behavior of our Azure resources—for instance, some of the common metrics are CPU and disk consumption and network latency.

Now, let's learn about metrics in ADX. Almost every Azure resource now has a **Monitoring** submenu in its properties panel, as shown in the following screenshot:

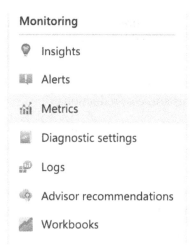

Figure 9.2 – Monitoring submenu

This is great, but as you deploy more and more Azure resources, you will find going to each resource to check metrics and logs is inefficient. Azure provides a centralized control blade for managing all your monitoring called **Azure Monitor**. **Azure Monitor** is beyond the scope of this book and deserves a book of its own.

By default, Azure enables basic metrics such as CPU consumption for our resources that are enabled by default. The ADX metrics are grouped into the following categories:

- **Cluster metrics** provides metrics related to the general health of our clusters—for example, CPU and cache utilization factor. One important note: if you are not using any autoscaling features, you should pay attention to the CPU and cache.

- **Ingestion metrics** provides metrics such as latency, blobs received, and jobs processed.

- **Streaming ingestion metrics** allows us to measure and observe the performance of our streaming operations.

- **Query metrics** allows us to measure and observe the performance of our queries. We will explore these metrics more in *Chapter 11, Performing Tuning in Azure Data Explorer.*

- **Export metrics** allows us to measure and observe the performance of our export operations—for example, failure and success rates.

- **Materialized views** allow is to expose an query over a table or another materialized view. Materialized views are useful for gaining insights for performance improvements and cost reductions.

Since ADX is regularly updated, I recommend visiting `https://docs.microsoft.com/en-us/azure/data-explorer/using-metrics` for the latest complete list of metrics that you can track.

Now that we know which metrics are available, the next step is to use them by plotting them on a graph. The **Metrics** option in the **Monitoring** submenu is where we can plot metrics, as shown in the following screenshot:

Figure 9.3 – Plotting ADX metrics

As shown in *Figure 9.3*, it is possible to plot multiple metrics on a graph. The following sequence of steps explains how to plot metrics on a graph and how to pin the graph to an Azure dashboard. Azure dashboards are the dashboards available in the Azure Portal, and not the dashboards available in the Data Explorer Web UI:

1. Log in to the Azure portal (`https://portal.azure.com`) and click on our ADX cluster.

2. In the ADX cluster's properties panel, scroll down to **Monitor** and click **Metrics**.

3. As shown in the following screenshot, select the **CPU** metric:

Figure 9.4 – Plotting ADX metrics (continued)

As shown in the following screenshot, the **CPU** metric, or whichever metric you choose, are immediately rendered on the screen:

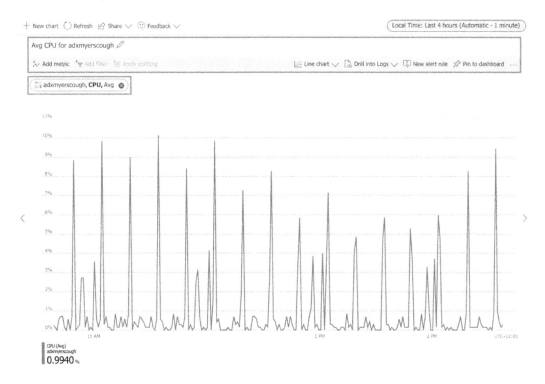

Figure 9.5 – Rendering the CPU metric

As shown in *Figure 9.5*, you can customize the title of a chart, add additional metrics to a chart, pin a chart to a dashboard, and generate alerts. Dashboards are a useful way of pinning charts from various sources and sharing them with colleagues. Dashboards with monitoring charts can also be used as your Azure portal home screen—for example, the following screenshot shows my overview dashboard, where I keep track of my ADX clusters, **Azure Active Directory (Azure AD)** users, and my monthly Azure costs:

Figure 9.6 – Azure dashboards

Although the dashboards are useful for displaying metrics and so on, there is a better alternative called **workbooks**. Azure workbooks are more feature-rich and have additional widget options. One of the nice features of workbooks is workbook templates. Workbook templates can be shared with the community, and Microsoft provides some ready-made templates to help with laying out your metrics.

> **Note**
> You can find workbooks under the **Insights** option—not **Workbooks**—under your clusters' **Monitoring** submenu in the properties panel, as shown in *Figure 9.7*.

The following screenshot shows where to find workbooks:

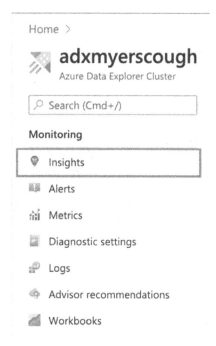

Figure 9.7 – Monitoring | Insights option with ADX clusters

In the next section, *ADX diagnostics*, we are going to discover how to enable logging, write logs to a Log Analytics workspace, and learn about the different tables that are available to us.

ADX diagnostics

Metrics are useful, but when you want more information, then you need to enable **diagnostics**. **Diagnostic settings** allows us to enable logging, provides detailed table-level information such as failed ingestion error codes and descriptions, as well as details on which queries have been executed on our clusters and databases.

By default, the diagnostic logs are disabled. I recommend enabling the diagnostic logs, especially in production environments. Note that you are charged by the amount of data you store in Log Analytics and storage accounts. Without the visibility provided by telemetry, it will be near impossible to know what is happening in your environments.

Enabling diagnostics is straightforward and you have the option to store them in Log Analytics (which is the preferred method), to a storage account, or an event hub. Before we can send our logs to a Log Analytics workspace, we need to create one.

Enabling Azure resource providers

Before we demonstrate how to create a Log Analytics workspace, it's important to understand the concept of **Azure resource providers**. Every Azure resource, such as VMs, Log Analytics workspaces, and ADX, all have a resource provider. Resource providers provide information regarding specific resources, such as the Log Analytics workspace, to **Azure Resource Manager** (**ARM**). The intention is to decouple resources from ARM. This design allows ARM to interact with resources without explicitly knowing all the details of a particular resource, such as the Log Analytics workspace, and allows third parties to integrate services such as **SendGrid**, which is a popular email service available on Azure. Not all Azure resource providers are enabled by default, and if you cannot find Log Analytics workspaces, then the chances are the resource provider is not enabled.

I have noticed that the resource provider is automatically enabled when deploying a Log Analytics workspace for the first time, but there were some instances where the resource provider was not enabled and the deployment failed. Resource providers are enabled at the *subscription scope*. The following sequence of steps explains how to enable the Log Analytics workspace provider in the Azure portal:

1. Log in to the Azure portal (`https://portal.azure.com`).

2. Click on your subscription. For example, my subscription is called `myerscough-adx-book`.

3. Under the subscription's properties panel, scroll down to **Settings** and click **Resource providers**.

4. Search for `Microsoft.OperationalInsights` and click **Register**, as shown in the following screenshot (you can skip this step if the resource provider is already enabled):

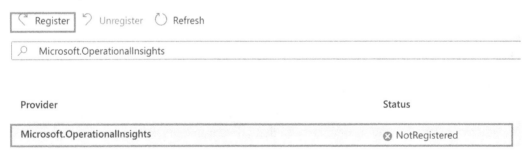

Figure 9.8 – Enabling resource providers

Once you have enabled `Microsoft.OperationalInsights`, you should be able to see **Log Analytics workspaces** in the portal.

Creating a Log Analytics workspace

The following sequence of steps explains how to create a workspace via the Azure portal:

1. Log in to the Azure portal (`https://portal.azure.com`).

2. Go to **All services** and search for `Log Analytics workspaces`.

3. Click **Log Analytics workspaces** and then click **+ Create**.

4. Ensure you have your subscription selected, and for the **Resource group** option, use the resource group where you deployed your ADX cluster. This is not a requirement; if you wanted to, you could deploy the Log Analytics workspace into a new resource group.

5. Give your workspace a name and select the same region you deployed your ADX cluster to. For example, I deployed my cluster to West Europe, so I selected **West Europe**.

6. Click **Next: Pricing tier >** and select **Pay-as-you-go (Per GB 2018)**.

7. We can skip the **Tags** section since we do not need tags for this example. Click **Review + create**.

8. Click **Create** once Azure completes its validation process, which usually takes a couple of seconds.

Now that we have a Log Analytics workspace, the next step is to configure and enable diagnostic logs.

Enabling diagnostic logs

The following sequence of steps demonstrates how to enable diagnostic logs and send them to the Log Analytics workspace we just created:

1. In the Azure portal, select your ADX cluster and click on the **Diagnostic settings** option in the ADX cluster's properties panel.

2. As shown in the following screenshot, click **+ Add diagnostic setting**:

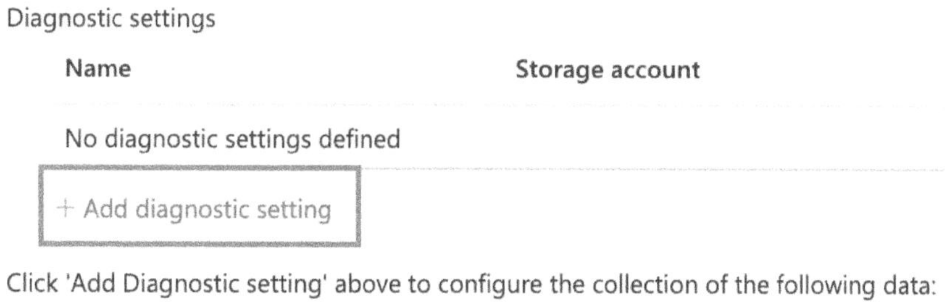

Figure 9.9 – Enabling diagnostics

3. Give your diagnostic setting a name, such as `adxDiagnostics`, and check all the **log** options, as shown in the following screenshot. This will enable all logging:

Figure 9.10 – Diagnostic configuration

4. Check **Send to Log Analytics workspace** and select the workspace you created earlier.

5. Finally, click **Save** to complete the process.

Based on my experience, it can take up to 15 minutes before data is generated and sent to your workspace, but once it is, you can go to the **Logs** option in your ADX cluster's properties panel and access the query editor. In the next section, *Troubleshooting ADX*, we will demonstrate how to query the logs when we troubleshoot an ingestion error.

Before moving on to the troubleshooting section, I wanted to give a quick overview of the tables that are available to us now that we have logging enabled, as follows:

- `SucceededIngestion` provides details on successful ingestion operations.

- `FailedIngestion` provides details on failed ingestion operations. We will use this table in the next section.

- `ADXIngestionBatching` provides details on batch operations, such as processing times and batch size.

- `ADXCommand` provides a list of commands executed on the ADX cluster. This table is useful for understanding which commands have been executed on your cluster and by whom.

- `ADXQuery` provides details on query execution. This table is useful when troubleshooting performance issues.

- `ADXTableUsageStatistics` provides basic statistical information regarding your tables.

- `ADXTableDetails` is a useful table that provides details regarding your database tables, such as retention policy, cache size, and cache policies.

- `AllMetrics` is all the metrics we were using in the previous section. When you enable **Diagnostics**, not only can you enable logging, but you can also export the metrics to another Log Analytics workspace, storage account, or event hub. When you have multiple Azure resources in your environment, it is a good practice to design a Log Analytics workspace strategy.

- `ADXJournal` contains information about metadata operations that are done on the ADX database.

In large environments, this is important to help scale and manage your metrics and costs. Log analytics can quickly grow since you are charged by the amount of storage you use, and I have seen some large bills just because teams had their own copies of data in their own Log Analytics workspaces.

Once you have your metrics and logs, the next step is to define alerts that notify you of issues.

Alerting in Azure

Another important aspect of monitoring is alerting. I am pretty sure you have more important things to do other than staring at monitoring dashboards. Instead, we can configure alerts so that we are notified when issues arise. There are two important components of alerts in Azure, as outlined here:

1. **Alert Rules** is where we define the scope of a rule. Rather than defining alerts to individual resources such as VMs, we can assign rules to a resource group or even a subscription. This means all VMs in a resource group or subscription will have an alert rule assigned, which is very useful when you manage large environments. An alert rule also comprises alert conditions. Alert conditions are expressions that should evaluate as either true or false. The expression can be based on metrics or logs.

2. **Action Groups** is where we configure which actions and **notifications** should be invoked when an alert condition evaluates as true. For example, we can have simple notifications that are email-based, and we can have more sophisticated actions such as executing an **Azure function** or **logic app**.

In the next section, *Troubleshooting ADX*, we will demonstrate how to troubleshoot ADX issues and how to configure email and **Short Message Service** (**SMS**) alerts so that we are notified when issues occur.

Troubleshooting ADX

As you may recall from *Chapter 4*, *Ingesting Data in Azure Data Explorer*, we set up infrastructure to ingest data from a storage account using an **event grid** and an **event hub**. Since we did not configure diagnostics at the time, the only way to check whether the ingestion succeeded was to run a query to check whether any data was available. Depending on the ingestion policy, you had to wait up to 5 minutes for the data to be ingested. Now, imagine an error occurred—how would you know? Should you refresh your browser or continuously execute a query to return the number of rows? No! That does not scale and, like me, you probably have better things to do with your time than continuously hitting *Shift + Enter* to execute a query.

In this section, we will intentionally introduce an error with our data ingestion process, and then we will learn how to troubleshoot such issues by looking at ADX's metrics and diagnostic logs using Log Analytics.

> **Note**
>
> In this section, we will reuse the infrastructure we created in *Chapter 4,*
> *Ingesting Data in Azure Data Explorer.* If you deleted the resources or did not
> complete the demonstration, please go back to *Chapter 4* and redeploy the
> infrastructure before proceeding.

From this point on, I will assume that you have all the necessary infrastructure deployed
and the ingestion process is working. In the next section, *Creating a new data connection*,
we will delete the existing data connection and create a new one.

Creating a new data connection

In this section, we will delete the existing data connection and create a new one, which
is almost identical to the existing data connection, except we will set the **Data format**
field to JSON instead of MULTILINE JSON. We could theoretically use the existing data
connection and just update the data format, but creating a new one is good practice for
your muscle memory.

The following sequence of steps explains how to delete an existing data connection from
an ADX database:

1. Log in to the Azure portal (https://portal.azure.com) if you are not
    already logged in.

2. Click on your ADX instance, and from your ADX instance's properties panel, under
    Data, click **Databases**. You should see the name of your ADX database in the list of
    available databases.

3. Click on your database, as shown in the following screenshot:

Figure 9.11 – Selecting your database

4. Click **Data connections**, which is under **Settings** in your ADX database's property panel. You should now see the data connection you created in *Chapter 4, Ingesting Data in Azure Data Explorer*, as shown in the following screenshot:

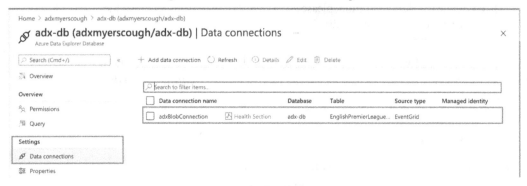

Figure 9.12 – ADX database's data connection

5. Select your data connection and click **Delete**. This will delete your data connection and you can track the deletion progress in the **Notifications** blade, as illustrated in the following screenshot:

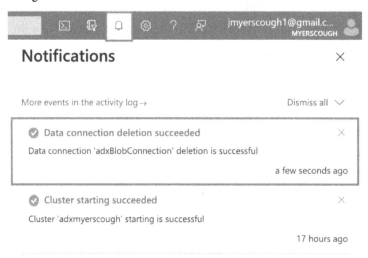

Figure 9.13 – Azure Notifications blade

Now that we have deleted our data connection, we need to create a new one. The following steps demonstrate how to create a new data connection and set the data format to JSON instead of MULTILINE JSON:

1. In your ADX database's properties panel, click **Data connections** and click **+ Add data connection**.

2. Select Blob storage as the connection type.

3. Give the data connection a name, such as `adxDataConnection`.

4. Ensure your subscription is selected and set the **Storage account** field to the account you created in *Chapter 4*, *Ingesting Data in Azure Data Explorer*.

5. Set the **Event type** field to `Blob created` and select **Manual** for the **Resources creation** setting.

6. Set the **Event Grid** field to the event grid you created in *Chapter 4*, *Ingesting Data in Azure Data Explorer*—for example, `adxEventGridSubscription`. The **Event Hub name** value will automatically be set.

7. Set the **Consumer group** field to `$Default`.

8. Click **Next: Ingest properties >** to configure the ADX database table and schema mapping, as shown in the following screenshot:

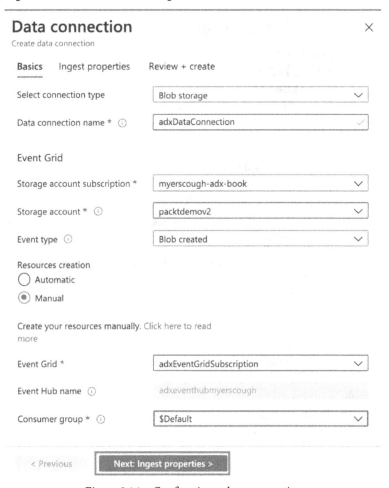

Figure 9.14 – Configuring a data connection

9. Enter the name of the English Premier League table we created in
    Chapter 4, *Ingesting Data in Azure Data Explorer*—for example,
    `EnglishPremierLeagueJSON`.

10. Set **Data format** to `JSON` and set the **Mapping name** field to the schema map
    we created in *Chapter 4*, *Ingesting Data in Azure Data Explorer*—for example,
    `EPL_Custom_JSON_Mapping`.

11. Click **Next: Review + create** > to review and validate our settings. Once the
    validation is complete, click **Create** to complete the creation process. The
    deployment progress can be tracked in the **Notifications** panel.

Now that we have set up our data connection (intentionally with an error), we will attempt
to ingest data in the next section, *Ingesting data to simulate an error*.

Ingesting data to simulate an error

In this section, we will upload one of the Premier League **JavaScript Object Notation**
(**JSON**) files (such as `${HOME}/Scalable-Data-Analytics-with-Azure-Data-Explorer/Chapter04/datasets/premierleague/json/season-1516_json.json`) to our storage account, which will trigger a `BlobCreated` event and
trigger our ingestion process. Since the data connection's data format is misconfigured,
the ingestion will fail.

The following steps explain how to upload a file to the storage account:

1. In the Azure portal, click on your storage account—for example, `packtdemo`.

2. In the storage account's properties pane, click **Containers**, which is located under
    the **Data storage** subheading.

3. Click the results container and then click **Upload** to upload a JSON file—for
    example, `${HOME}/Scalable-Data-Analytics-with-Azure-Data-Explorer/Chapter04/datasets/premierleague/json/season-1516_json.json`.

This is a perfect example of why we need to monitor our resources, not just in Azure but
in general. Without monitoring and alerting in place, we have no idea if an operation is
going to fail or succeed without actively checking the system/database for new data. Keep
in mind this example is supposed to fail and can take up to 5 minutes to complete due
to the ingestion policy. In the next section, *Observing and troubleshooting ADX*, we will
manually observe the ingestion process and then troubleshoot and determine why the
ingestion is failing.

Observing and troubleshooting ADX

In this section, we will manually monitor our ADX cluster to check our JSON data has been ingested, and when you see that it has not been ingested, we will then troubleshoot the issue by applying the method of breaking down problems through the process of elimination.

If we look at our Event Hubs metrics, we can see that a message was received and sent, as shown in the following screenshot:

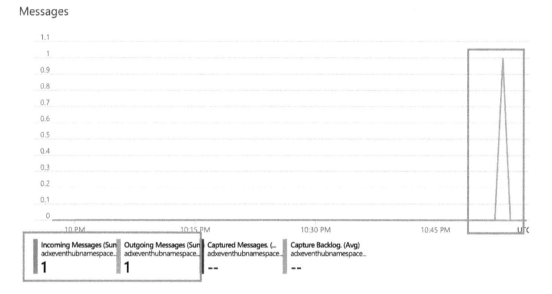

Figure 9.15 – Event Hubs processing the Blob events

We'll now head back over to the Data Explorer web user interface (UI) and query our EnglishPremierLeagueJSON table, for example, by running the following code:

```
EnglishPremierLeagueJSON
| count
```

We will see that no data has been ingested. Now, imagine the scenario: you are a site reliability engineer, you are on call, and you receive a page in the middle of the night saying the user is not seeing any new data. As mentioned earlier in the chapter, the key to troubleshooting is to understand the issue or problem statement, which in this scenario is that users are not seeing new data, and to narrow down the problem as quickly as possible by eliminating potential issues. Ideally, you will have monitoring and alerting in place, which will help narrow down the issue.

So, let's break this down, as follows:

1. Understand the problem: *users cannot see new data*. The first thought that comes to my mind is whether the problem is on the server side or the client side. Since my primary responsibility is the server side, this is where I will focus.

2. We know our ADX instance is hosted on Azure. The next step in our elimination process is to rule out an issue with the Azure platform itself. As mentioned earlier in the chapter, we can use the **Service Health** service to eliminate any possible issues with the Azure platform. It is a good idea to not just look at your Azure resource, but also some of the fundamental services your Azure service— in our case, ADX— uses, such as **VMs**, **storage**, and **networking infrastructure**. In my experience, issues with these services can indirectly impact other services in Azure. As shown in the following screenshot, there are no health issues with the ADX cluster. I would recommend adding resource health alerts in a real environment:

Figure 9.16 – Service health status

3. Now that we have ruled out the possibility of an issue with Azure, we can focus our attention on our ADX cluster. The next step is to look at our ADX cluster's **Insights** page. When I first started writing this book, the **Insights** workbook was not available. Over the course of writing this book, Microsoft has extended the workbook and added some insights regarding ingestion. As shown in the following screenshot, we can immediately see there is indeed a **Failed Ingestions** instance, and in keeping with the process of elimination, we can focus on the ingestion aspects. In my experience, in cases such as this, do not be too quick to completely rule out other issues. This failed ingestion could be a symptom of another issue. I have experienced enough highs and lows thinking I had found the root cause, only to discover it was a symptom of the real problem:

Figure 9.17 – ADX cluster's Insights workbook

4. As shown in *Figure 9.17*, there is a tab called **Ingestion (preview)**, which shows a counter for the successful and failed ingestion attempts. Let's dig deeper by looking into the diagnostic logs to understand what exactly is causing the issue.

5. As mentioned in the earlier section, *Monitoring ADX*, I explained there are several log tables available that we are sending to Log Analytics. Since we know there is a failed ingestion, the table we are interested in is—aptly—called `FailedIngestion`.

6. In the Azure portal, click your ADX cluster, and in the properties panel, under **Monitoring**, click **Logs** to open the Log Analytics query editor. You will see that Microsoft provides a few useful predefined queries in the **Queries** pop-up dialog window.

7. If you run the following query, you should see at least one record:

```
FailedIngestion
| count
```

8. Run the following query to display details of the error:

```
FailedIngestion
| project TimeGenerated, OperationName, Category,
ResultType, Database, Table, ErrorCode, Details,
IngestionSourcePath
```

As described in *Chapter 4*, *Ingesting Data with Azure Data Explorer*, there is a difference between JSON and MULTILINE JSON. Earlier in the chapter, we created a new data connection and used JSON instead of MULTILINE JSON.

JSON expects each record in the JSON file to be on one line, whereas MULTILINE JSON—as the name implies—allows records to be multiline records. As shown in the following screenshot, the ErrorCode and Details columns inform us there is a problem with our JSON format:

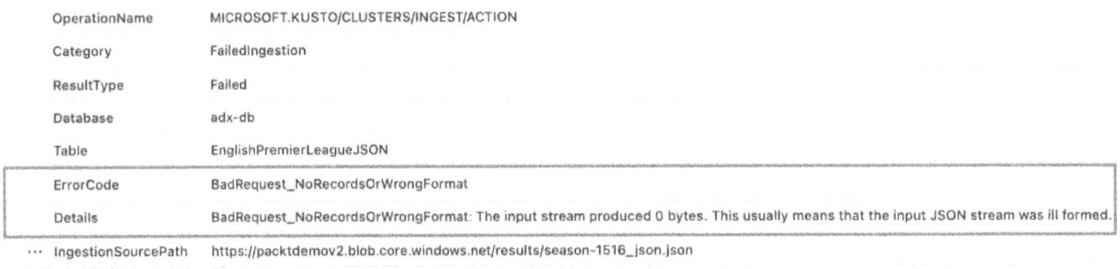

OperationName	MICROSOFT.KUSTO/CLUSTERS/INGEST/ACTION
Category	FailedIngestion
ResultType	Failed
Database	adx-db
Table	EnglishPremierLeagueJSON
ErrorCode	BadRequest_NoRecordsOrWrongFormat
Details	BadRequest_NoRecordsOrWrongFormat: The input stream produced 0 bytes. This usually means that the input JSON stream was ill formed.
⋯ IngestionSourcePath	https://packtdemov2.blob.core.windows.net/results/season-1516_json.json

Figure 9.18 – Finding the data ingestion root cause

Whenever there are issues in production environments, it is common practice to conduct a postmortem and review what happened and how we can prevent or reduce the risk of the issue reoccurring. In our case, one improvement we could make is to implement monitoring and alerting. Rather than refreshing our dashboards, it would be nice to be notified when the ingestion process fails. In the next section, *Configuring alerts for ingestion failures*, we will configure an alert to send an email and SMS message whenever data ingestion fails.

Configuring alerts for ingestion failures

In this section, we will demonstrate how to set up action groups and configure an alert to send both email and SMS notifications whenever data ingestion fails.

The following sequence of steps demonstrates how to set up and configure an alert that will notify us via email and SMS:

1. Log in to the Azure portal (`https://azure.portal.com`).

2. The next step is to create our action group. Select your ADX cluster, and in the properties pane, scroll down to **Monitoring**, and then click **Alerts**. Then, click **Manage actions**, as shown in the following screenshot:

Figure 9.19 – Creating a new action group

3. Click + **New action group**.

4. Ensure you have the correct subscription selected and select the resource group where your ADX cluster is deployed. The action group and ADX do not need to be in the same resource group—I am just putting them in the same resource group for convenience.

5. Give your action group a name—for example, `adxIngestionAlert`.

6. Give your action group a display name—for example, `adxIngestion`. Please note that the display name is limited to 12 characters at the time of writing.

7. Click **Next: Notifications >**.

8. For the **Notification type** setting, select `Email/SMS message/Push/Voice` and then check **Email** and **SMS** and enter your phone number and email address. Finally, click **OK**, as shown in the following screenshot:

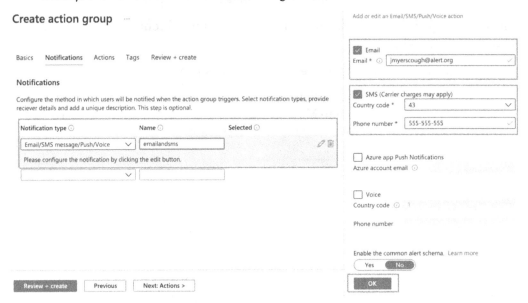

Figure 9.20 – Action group notification configuration

9. Next, click **Review + create**. Finally, click **Create** once the validation is complete.

    Our newly created action group will send an email to `jmyerscough@alert.org` and an SMS message to `+43 555-555-555`. Action groups are typically created immediately, but can take a couple of minutes to appear in your list of action groups.

10. As shown in the following screenshot, click your ADX cluster name to go back to the **ADX overview** blade:

Figure 9.21 – Azure portal's breadcrumb trail

11. Select your ADX cluster, and in the properties pane, scroll down to **Monitoring** and click **Logs**.

12. Next, we need to write a query that will count the number of failed ingestion log entries. Paste the following query and press the *Shift + Enter* keys to execute the query. You should see a count of at least 1 since we generated an ingestion error earlier:

```
FailedIngestion
| count
```

13. As shown in the following screenshot, click **+ New alert rule** to create a new alert based on our query:

Figure 9.22 – Creating a new alert

14. Ensure the search query is set to the following:

```
FailedIngestion
| count
```

15. Under **Measurement**, set **Measure** to Count and leave **Aggregation Type** at Total and **Aggregation granularity** at 5 minutes.

16. Under **Alert Logic**, keep **Operator** set to Greater than, set the **Threshold value** field to 0, and click **Next: Actions >**.

17. Select your action group—for example, adxIngestionAlert—and click **Next: Details >**.

18. It is important to categorize your alerts. For example, some alerts are normally critical and require immediate attention, whereas others could just be informational. In this example, set the **Severity** field to 0 – Critical. I would consider ingestion failures as critical since this would result in no data being available.

19. Give your alert a name—for example, Ingestion failure.

20. Enter a good description for your alert, such as alert when there are ingestion failures.

21. Ensure you have the correct subscription selected, select the resource group where your ADX is deployed (such as adx-rg), and set the region to where your ADX cluster is.

22. Next, click **Review + create**. Finally, click **Create** once the validation is complete.

23. Now that the alert is configured, upload a JSON file such as `${HOME}/Scalable-Data-Analytics-with-Azure-Data-Explorer/Chapter04/datasets/premierleague/json/season-1617_json.json`. Depending on the ingestion policy, the workflow can take up to 10 minutes.

Since we know the ingestion will fail, you will receive an alert via email and SMS notifying you of the issue, as shown in the following screenshot:

Figure 9.23 – Ingestion failure alert

In this section, we covered how to troubleshoot ADX issues. Teaching troubleshooting is a difficult task; it is not possible to teach troubleshooting solutions—instead, we must learn and master the troubleshooting process by process of elimination. This book is not intended to be a course on troubleshooting, but I hope I provided some insights. The best way to learn to troubleshoot is to practice and gain experience.

One of my long-term mentors, Arunee Singhchawla, taught me to emphasize a couple of key concepts that have helped me sharpen my skills, as follows:

- Include monitoring in your initial requirements and analysis; do not treat monitoring as an afterthought or a nice-to-have feature.

- Clearly define your SLIs, SLOs, and SLAs.

- Set up regular troubleshooting drills. As the saying goes, practice makes perfect.

Without monitoring and alerting, troubleshooting becomes an almost impossible task.

Summary

We have only scratched the surface with regard to monitoring and troubleshooting. Monitoring and Azure Monitor deserve their own book in order to do them any justice.

In this chapter, I began by introducing the concept of monitoring, discussing why monitoring is important and what SLIs, SLOs, and SLAs are. Then, I introduced the concept of troubleshooting and discussed my thought process and how I break down problems.

The rest of the book then focused on the key metrics and logs available to us for ADX and demonstrated how to enable diagnostics, and then we walked through an example and troubleshot an issue where data ingestion was not working.

Finally, we learned how to configure alerts for ingestion failures. We configured an action group that would send an email and an SMS whenever the ingestion failed.

In *Chapter 10, Azure Data Explorer Security*, we will learn how to secure our ADX clusters.

Questions

Before moving on to the next chapter, test your knowledge by answering the following questions. The answers can be found at the back of the book:

1. What is the difference between SLIs, SLOs, and SLAs?
2. Configure a metrics dashboard to display the **Blobs received** metric and then import some data into your ADX cluster. What do you see?
3. Try to implement monitoring alerts for an event hub's incoming and outgoing messages and set the severity to **Informational** since this is typically not an error.
4. How many severity levels are there and what does each level mean?

10

Azure Data Explorer Security

As recently as 6 years ago, the emphasis on **public cloud security** was not what it is today. I remember that, at one point, disk encryption was not available on storage accounts. Then, it became an option that could be enabled, and today, it is enabled by default.

One of the biggest concerns with the public cloud is ensuring that our data and resources are not accessible to just about anyone on the internet. In this chapter, we will learn how to secure our **Azure Data Explorer** (**ADX**) instances using **identity management** and how to perform **network filtering** on unwanted traffic and bad actors.

We will begin by introducing some of the basic terminology and concepts you should be familiar with, such as the principle of least privilege and **role-based access control** (**RBAC**). Next, we will explore the concepts of identity management with **Azure Active Directory** (**AAD**). We will also learn about the differences between security principals, users, groups, and service principals and look at the different levels of access to our resources; for example, the management plane versus the data plane. After understanding the theoretical background, we will learn how to assign permissions to ADX clusters, databases, and tables using the **Azure portal** and **KQL** management commands.

Next, we will understand how to restrict access to our ADX cluster using virtual networks and subnet delegation.

Finally, we will learn about **network security groups** (**NSGs**) and how to use them to filter both inbound and outbound traffic. We will also learn how to enable diagnostic logging so that we can observe network traffic being accepted and dropped. In the practical example, we will learn how to restrict the incoming traffic from our public IP.

In this chapter, we will cover the following main topics:

- Introducing identity management
- Introducing virtual networking and subnet delegation
- Filtering traffic with NSGs

Technical requirements

The code examples for this chapter can be found in the `Chapter10` folder of this book's GitHub repository: `https://github.com/PacktPublishing/Scalable-Data-Analytics-with-Azure-Data-Explorer.git`.

Introducing identity management

When you signed up to Azure earlier in this book, an AAD tenant was created. AAD is Microsoft's cloud-based identity and access management service and is used by other major services such as **Office365**. AAD is a great service that allows you to manage your users and devices, supports **multi-factor authentication** (**MFA**) and **privileged identity management** (**PIM**), and so on. I would not be doing AAD justice if I tried to cover everything about AAD here. If you would like to learn more, I recommend Packt Publishing's *Mastering Identity and Access Management with Microsoft Azure – Second Edition*.

Before learning how to manage users, it is important to understand RBAC and the differences between the management plane and the data plane.

Introducing RBAC and the management and data planes

Before we introduce RBAC, it is important to understand what **authentication** and **authorization** are and how they differ from one another. **Authentication** is how we prove/authenticate someone's identity, while **authorization** is related to giving someone access to our resources. For example, when we log into the Azure portal, we authenticate by entering our credentials. Once logged in, we will see the resources we are authorized to see.

Another important concept in cyber security is the **principle of least privilege** (**PoLP**). This is where you assign a user/security principal enough permission that they can do their job and nothing more. Azure simplifies implementing PoLP with its advanced identity management features, such as AAD and RBAC.

AAD uses objects known as **security principals** to store users, groups, and service principals (identities for applications). These objects can be authorized to give you access to your Azure resources, such as ADX clusters, Log Analytics workspaces, and virtual networks via Azure RBAC.

Azure RBAC allows us to apply the PoLP and manage access to our resources. In other words, RBAC allows us to define who is authorized to do what at a given scope.

A role in Azure is a collection of actions/permissions that are either allowed or denied. Roles are implemented in JSON and unless you are writing custom roles, you will not need to work directly with these JSON files. The following screenshot depicts the JSON definition for the `Contributor` role. The `Contributor` role authorizes a user to do everything except manage access. As shown on *line 12* of `Chapter10/rbac/contributor.json`, the asterisks denote all actions. On *line 14*, `notActions` declares the actions that are not permitted by the contributor. For instance, read and write actions are disabled for authorization operations (`Microsoft.Authorization/*`):

```
Chapter10 > rbac > {} contributor.json > {} properties > [ ] permissions > {} 0 > [ ] actions > ▥ 0
 1    {
 2          "id": "/providers/Microsoft.Authorization/roleDefinitions/b24988ac-6180
 3          "properties": {
 4              "roleName": "Contributor",
 5              "description": "Grants full access to manage all resources, but doe
 6              "assignableScopes": [
 7                  "/"
 8              ],
 9              "permissions": [
10                  {
11                      "actions": [
12                          "*"
13                      ],
14                      "notActions": [
15                          "Microsoft.Authorization/*/Delete",
16                          "Microsoft.Authorization/*/Write",
17                          "Microsoft.Authorization/elevateAccess/Action",
18                          "Microsoft.Blueprint/blueprintAssignments/write",
19                          "Microsoft.Blueprint/blueprintAssignments/delete",
20                          "Microsoft.Compute/galleries/share/action"
21                      ],
22                      "dataActions": [],
23                      "notDataActions": []
24                  }
25              ]
```

Figure 10.1 – RBAC role definition

At the time of writing, Azure provides over 200 built-in roles and supports custom roles. The most common and general built-in roles are as follows:

- Owner: Authorizes security principals to perform read and write operations on all resources and authorizes the ability to perform role assignments
- Contributor: Authorizes security principals to perform read and write operations on resources and does not authorize security principals to perform user assignment actions
- Reader: Authorizes security principals to perform read actions

As we mentioned earlier, role assignments can be made at different scope levels and take advantage of the parent-child relationship between management groups, subscriptions, resource groups, and resources, which means children inherit the permissions from their parents.

The various scope levels with regards to RBAC assignment are as follows:

- Management groups: A logical container for managing your subscriptions.
- Subscriptions: This consists of resource groups and Azure resources such as virtual networks.
- Resource groups: A logical container for Azure resources.
- Resources: Azure resources such as ADX.

It is not recommended to perform role assignments at the resource scope as this can become difficult to manage due to the increasing number of Azure resources. Depending on the size of your environment, assigning permissions at the highest level of scope makes sense if you wish to take advantage of the parent-child inheritance benefits. For example, if we assign the Contributor role at the Subscription scope, those permissions will be inherited by all the resource groups and resources in the subscription.

RBAC role assignments are evaluated by **Azure Resource Manager** (**ARM**). As we mentioned in *Chapter 2*, *Building Your Azure Data Explorer Environment*, Azure provides a service known as ARM, which is the deployment and management service that allows us to create, update, and delete our resources. The following diagram depicts the RBAC evaluation process:

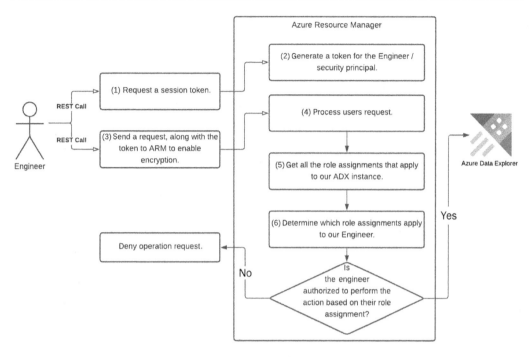

Figure 10.2 – RBAC evaluation process

When a user authenticates with ARM, they are assigned a token that is passed with all the other subsequent REST calls to ARM.

> **Note**
>
> ARM exposes a REST API, which is used by clients such as the Azure portal to manage Azure resources. **Representational State Transfer (REST)** is a popular software architecture particularly in web and microservice development. REST is based on HTTP methods such as GET, POST, and DELETE. I recommend Packt's *Hands-On RESTful Web Services with Go – Second Edition* for more information regarding REST.

In the preceding diagram, the engineer would like to enable encryption on his ADX cluster. The REST call is sent to ARM, which evaluates the RBAC assignments before either allowing or denying the operation.

It is also important to know there are two dimensions of access – the management plane and the data plane:

- The **management plane** refers to the permissions required to manage and configure your Azure resources, such as scaling the ADX cluster, configuring your NSGs, and so on. Users that consume and query your ADX cluster do not necessarily need access to the management plane to use your resources.

- The **data plane** refers to the permissions required to use and consume the resources. For example, to access and connect to ADX clusters, you need permissions set at the database level.

Another benefit of using **RBAC** is that it can reduce the risk of accidental operations. I remember an incident when an engineer had more access than he needed due to everyone using the contributor role. He accidentally reassigned a subscription to another Azure AD tenant, which resulted in all the RBACs assigned to the subscription being deleted. This meant nobody was able to access or see the subscription due to the permissions being deleted. To avoid such problems, ensure you apply the PoLP and only grant enough access for someone to do their job and nothing more. For instance, if an engineer's responsibility is to manage virtual machines, then grant them the `Virtual Machine Contributor` role rather than the general `Contributor` role.

In the next section, we will learn how to grant access to the management plane so that users can help support and manage our ADX infrastructure.

Granting access to the management plane

At the time of writing, there are approximately 295 built-in Azure roles, and the list is growing as Microsoft adds more and more resources to the Azure platform. These roles can be grouped into the following categories:

Role Category	Description
General	Roles that are not related to a specific Azure resource type. Later in this chapter, we will demonstrate how to assign the `Reader` and `Contributor` roles.
Compute	Roles related to virtual machine management.
Networking	Roles related to virtual network management.
Storage	Roles related to storage management.
Web	Roles related to the app server and web (media and search) management.

Role Category	Description
Containers	Roles related to container and Kubernetes management.
Databases	Roles related to database management. Interestingly, there are currently no roles specific to ADX.
Analytics	Roles related to analytics management. Interestingly, there are currently no roles specific to ADX.
Blockchain	Roles related to blockchain management.
AI + machine learning	Roles related to AI management.
Internet of things	Roles related to IoT management.
Mixed reality	Roles related to mixed reality management, such as augmented reality and virtual reality.
Integration	Roles related to API management.
Identity	Roles related to user and group management.
Security	Roles related to security management, such as Azure Sentinel and Azure Key Vault.
DevOps	Roles related to DevTestLab management.
Monitor	Roles related to monitoring management.
Management + governance	Roles related to resources such as cost management and automation.
Other	A group of miscellaneous roles.

Table 10.1 – Azure RBAC categories

A complete list of roles can be found at `https://docs.microsoft.com/en-us/azure/role-based-access-control/built-in-roles`. I highly recommend that you take your time to learn about the different roles and apply the principle of least privilege when designing an access policy for your environments. The list of available roles can also be obtained via PowerShell using the following cmdlet:

```
Get-AzRoleDefinition | Select-Object Name, Description
```

Now that you are familiar with the different Azure RBAC roles, we will demonstrate how to add access to the management plane.

> **Note**
>
> Access can be assigned at different levels of scope within Azure. It is possible to assign access at the resource instance level, such as directly on the ADX cluster, the resource group, subscription, management group, and so on.
>
> When the role is assigned at the resource group level, the user would have access to everything in the resource group. The same applies to subscription-level access.
>
> If a role is assigned at the subscription level, the access would be for everything in the subscription. As I mentioned earlier, I recommend Packt's *Mastering Identity and Access Management with Microsoft Azure – Second Edition*, if you are interested in learning more about identity management on Azure.

The following steps demonstrate how to assign the Reader role at the subscription level to a user called Harrison:

1. Open the Azure portal (`https://portal.azure.com`) and click the subscriptions icon.

2. If you do not have subscriptions pinned to your sidebar, click **All services** and search for `subscriptions`.

3. Click **Subscriptions** and then click your subscription; for example, `myerscough-adx-book`.

4. In the **Properties** pane, click **Access control (IAM)**.

5. Click **+ Add** and then **Add role assignment**, as shown in the following screenshot:

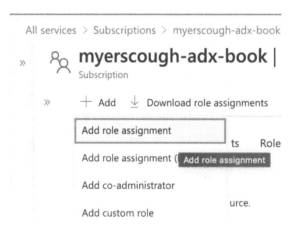

Figure 10.3 – Adding a new role assignment

6. On the **Add role assignment** blade, select the **Reader** role from the **Role** drop-down menu.

7. Set **Assign access to** to User, group, or service principal.

8. In the **Select** text box, search for your user – for example, harrison – and click on the user record.

9. Click **Save** to add the user, as shown in the following screenshot:

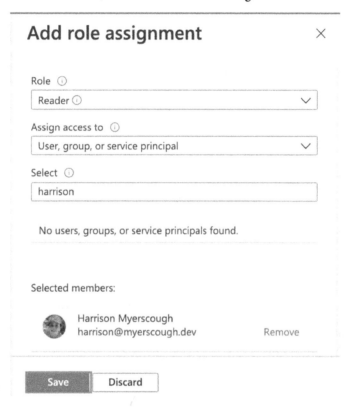

Figure 10.4 – Assigning the Reader role to a new user

We can verify that the role has been assigned by clicking on **Role assignments**. The **Role assignments** tab shows a list of users, along with their assignments:

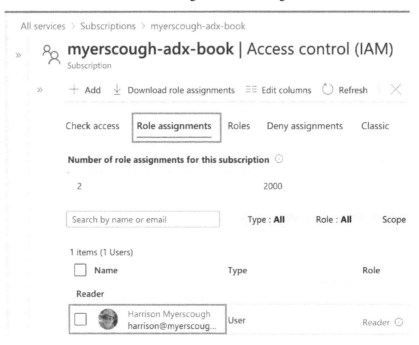

Figure 10.5 – Verifying role assignments

We can also verify read-only access by logging in as Harrison and trying to update the cluster SKU. When I tried to modify the cluster when signed in as Harrison, I got the following error:

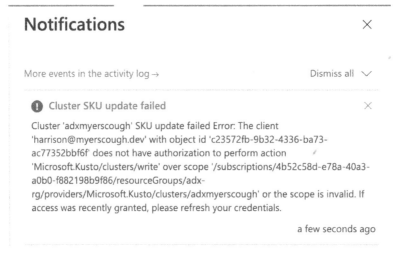

Figure 10.6 – Verifying reader access

In this section, we learned how to assign access to the management plane via the Azure portal. As we mentioned previously, I highly recommend that you apply the PoLP and only assign the access required, nothing more.

In the next section, we will learn how to add users to the database at the data plane level.

Granting access to the data plane

Like the management plane roles, some roles are predefined by Microsoft for cluster, database, and table access. The list of database and table roles is a lot smaller than the management plane roles. The following table provides an overview of the roles that are currently available at the time of writing:

Role	Description
AllDatabasesAdmin	Administration access for all databases. This permission is assigned to the cluster scope.
Admin	Administration access for specific databases. This permission is assigned at the database scope level.
User	Can create tables, functions, and so on within a database. This permission is assigned at the database scope level.
AllDatabasesViewer	Can read all data and metadata for all databases. This permission is assigned to the cluster scope level.
Ingestor	Can ingest data into tables but cannot query data. This permission is assigned at the database scope level.
UnrestrictedViewer	Can query all the tables that have a restricted access policy assigned. This permission is assigned at the database scope level.
Monitor	Can execute .show commands. This permission is assigned at the database scope level.
Table admin	Admin of a particular table. Table access can only be granted using KQL management commands at the time of writing. This permission is assigned at the table scope level.
Table ingestor	Can ingest data into a table but cannot query the data. Table access can only be granted using KQL management commands at the time of writing. This permission is assigned at the table scope level.

Table 10.2 – Database and table roles

Now that we have an overview of the different roles, let's explore how to add a user at the data plane scope level by assigning the `AllDatabasesAdmin` role to `James`.

Adding permissions via the Azure portal is straightforward. The following steps show how to add an administrator (`AllDatabasesAdmin`) permission to our ADX cluster:

1. Go to `https://portal.azure.com` and click on your ADX cluster; for example, `adxMyerscough`.

2. Under your cluster's properties, click **Permissions**, as shown in the following screenshot:

Figure 10.7 – Permissions at the data plane level

3. Click **+ Add**. Then, from the drop-down menu, select **AllDatabasesAdmin**.

4. Search for a user in your AAD. For example, I am going to make a user called `James` one of the database administrators.

5. Click **Select** once you have selected your user to complete this process. Now, you should see the user listed as an admin, as shown in the following screenshot:

Name	Type	Role	Tenant Name
James Myerscough james@myerscough.dev	User	Cluster AllDatabasesAdmin ⓘ	Myerscough

Figure 10.8 – Adding a new administrator

To assign permissions at the database scope level, click one of your databases. Then, under the **Properties** blade, select **Permissions**, as shown in the following screenshot. The steps to assign a user or group are the same as those described previously:

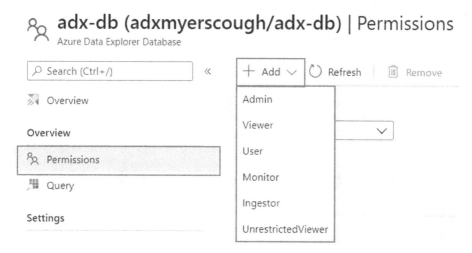

Figure 10.9 – Database scope permissions

In this section, you learned how to grant users access to the data plane scope and assigned the AllDatabasesAdmin role to a user. In an environment where you have a team, it is important to assign your team members with the correct level of access. Remember to apply the principle of least privilege.

As you may recall from when we deployed our ADX cluster, in *Chapter 2, Building Your Azure Data Explorer Environment*, under the **Security** tab, there were configuration options regarding identity management. Although we will not be using managed identity in this book, it is worth mentioning that ADX does support **Azure Managed Identity**. As your cluster grows and you add integrations such as custom applications, you should consider managing identities and authentication. With this, one common challenge and potential security risk is managing secrets and credentials. Azure solves this problem using the concept of managed identity. With managed identity, Azure takes care of managing the passwords both in terms of storing them and rotating them. If you are interested, more information regarding managed identity can be found here: https://docs. microsoft.com/en-us/azure/active-directory/managed-identities-azure-resources/.

Before moving on, I would like to point out, ADX supports both system-assigned identity and user-assigned identity. The key difference between them is system-assigned identity has the same life cycle as your cluster. This means, when you delete the cluster, the identity will also be deleted. Clusters can only have one system-assigned identity assigned at a given time. User-assigned identities have a different life cycle to the cluster. It means that they are not deleted when your cluster is deleted. Clusters can have multiple user-assigned identities assigned at a given time. My recommendation would be to use user-assigned identities for your custom applications. Please note, it is bad practice to share credentials/identities.

In the next section, we will discover how to secure our ADX clusters by deploying them in a virtual network that allows us to restrict connectivity to the cluster.

Introducing virtual networking and subnet delegation

As we saw in the previous section, *identity management* is a good method for restricting access to ADX clusters. We can control access at both the management and data plane levels, but our cluster is still available on the public internet. Anyone who knows the name of our cluster could potentially connect by guessing usernames and passwords.

Like a lot of Azure resources, such as storage accounts and Azure SQL, they are accessible on the internet by default. The problem with this default deployment is that we cannot restrict *inbound and outbound traffic*. Azure supports advanced deployments that allow us to deploy resources within a virtual network. Virtual networks let us create private networks on Azure to isolate and restrict access to our resources, such as virtual machines, ADX clusters, and so on.

Deploying our ADX cluster in a virtual network gives us more control over inbound and outbound traffic. We can use NSGs to restrict traffic and we can use virtual appliances such as Azure Firewall for more advanced features.

As shown in the following diagram, the secure deployment is more involved and requires additional Azure resources, which will we discuss later in this chapter. The biggest difference to note in the following diagram is the connectivity. In the default deployment, all connectivity is via the internet, whereas in the secure deployment, we can restrict connectivity to VPN, ExpressRoute, or specific IP addresses on the internet:

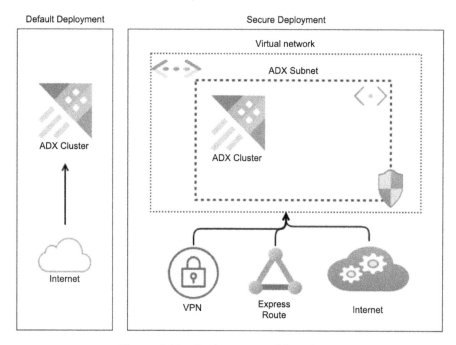

Figure 10.10 – Deployment models in Azure

The remainder of this chapter will explain and demonstrate how to deploy an ADX cluster in a virtual network, restrict internet access, and only allow traffic from your public IP address.

As we mentioned earlier, there are several Azure resources we need to deploy before we can deploy our ADX cluster in a virtual network:

- The first resource we need is a **virtual network**. Virtual networks allow us to create private networks on Azure to isolate and restrict access to our resources. Every virtual network has an IPv4 address space, which is a pool of IP version 4 addresses that you can use within your network, such as 10.0.16.15. Virtual networks consist of one or more subnets. **Subnets** are the sub-networks in your virtual network where you deploy your resources such as virtual machines. In our example, we will only create one subnet, which we will use for deploying our ADX cluster. In production environments, it is common practice to create multiple subnets so that we can have granular control over communication within our virtual network.

- The second resource we need is an NSG. An NSG contains a group of rules that allow us to accept or deny inbound and outbound traffic. For example, we could restrict all virtual machines from accessing the public internet or we could restrict access to our ADX cluster to just our public IP address.

- The third resource we need is a **route table**. Route tables allow us to create **user-defined routes**, meaning we can force or route network traffic to the desired location. In this example, we will not be adding any user-defined routes. The ADX cluster, during its deployment, will add routes to ensure ADX cluster traffic is routed to our virtual network.

- The fourth item we need is *two* **public IP addresses**. As you know, ADX clusters have two public IP addresses – one for general connectivity and another that's used specifically for data ingestion. For example, `https://adxmyerscough.westeurope.kusto.windows.net` is used for general connectivity and `https://ingest-adxmyerscough.westeurope.kusto.windows.net` is used for data ingestion.

- The final resource is the **ADX cluster**. Once we have deployed the prerequisite infrastructure, we will deploy the ADX cluster to our virtual network.

In the next section, *Creating a new resource group*, we will create a new resource group where we will store all our resources related to the subnet delegation demonstration.

> **Note**
>
> We will deploy all our resources via the ARM templates, which can be found in the `Chapter10` directory in this book's GitHub repository. The deployment script, called `adx-arm-deploy.ps1`, can deploy all the infrastructure at once. In the spirit of learning, we will be executing each `New-AZResourceGroupDeployment` separately to understand the resource deployment in more detail. Like the examples in previous chapters, we will deploy our infrastructure from **Azure Cloud Shell**.

Creating a new resource group

Before we deploy any infrastructure, let's create a new resource group called `adx-subnet-delegation-rg` that will be used to store all our resources associated with this demo. Then, at the end, you can simply delete the resource group to clean up your environment:

1. Go to `https://shell.azure.com`. This will take you to the Azure portal but will open Azure Cloud Shell for you.

2. If you have been following the previous examples, navigate to the `Chapter10` directory by typing `${HOME}/Scalable-Data-Analytics-with-Azure-Data-Explorer/Chapter10`. Skip to *Step 5* if you have been following along. If you have not, please proceed with *Step 3*.

3. If you have not been following along, create a new directory called `development` by typing `mkdir development`. Then, navigate to the development directory by typing `cd development`.

4. Clone the GitHub repository by typing `git clone https://github.com/PacktPublishing/Scalable-Data-Analytics-with-Azure-Data-Explorer.git`. Then, navigate to `Chapter10` by typing `cd Scalable-Data-Analytics-with-Azure-Data-Explorer/Chapter10`.

5. Type `code  .` to open the lightweight Visual Studio code editor.

6. Click **adx-arm-deploy.ps1** to open the file in the editor, as shown in the following screenshot:

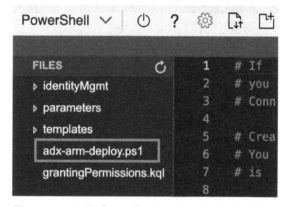

Figure 10.11 – Lightweight Visual Studio Code Editor

7. Copy and paste *line 9*, `$resourceGroupName = "adx-subnet-delegation-rg"`, into Cloud Shell and hit *Enter*. This will create a variable called `$resourceGroupName` that we can reuse with our cmdlets.

8. Copy and paste *line 10*, `$location = "westeurope"`, into Cloud Shell and hit *Enter*. Feel free to use a location closer to you. See *Chapter 2, Building Your Azure Data Explorer Environment*, to learn how to get a list of available Azure regions using PowerShell.

9. Copy and paste *line 11*, `New-AzResourceGroup -Name`
    `$resourceGroupName -Location $location -Force`, to create a new
    resource group. As shown in the following screenshot, `ProvisioningState` will
    indicate whether the operation succeeded or not:

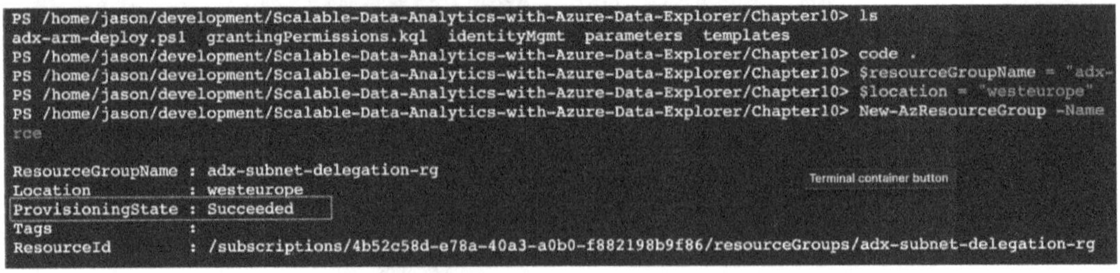

```
PS /home/jason/development/Scalable-Data-Analytics-with-Azure-Data-Explorer/Chapter10> ls
adx-arm-deploy.ps1  grantingPermissions.kql  identityMgmt  parameters  templates
PS /home/jason/development/Scalable-Data-Analytics-with-Azure-Data-Explorer/Chapter10> code .
PS /home/jason/development/Scalable-Data-Analytics-with-Azure-Data-Explorer/Chapter10> $resourceGroupName = "adx-
PS /home/jason/development/Scalable-Data-Analytics-with-Azure-Data-Explorer/Chapter10> $location = "westeurope"
PS /home/jason/development/Scalable-Data-Analytics-with-Azure-Data-Explorer/Chapter10> New-AzResourceGroup -Name
rce

ResourceGroupName : adx-subnet-delegation-rg
Location          : westeurope                                    Terminal container button
ProvisioningState : Succeeded
Tags              :
ResourceId        : /subscriptions/4b52c58d-e78a-40a3-a0b0-f882198b9f86/resourceGroups/adx-subnet-delegation-rg
```

Figure 10.12 – ProvisioningState

Now that we have created our resource group, `adx-subnet-delegation-rg`, we can
begin deploying our resources. In the next section, we will deploy our NSG and look at
the default inbound and outbound rules that are created.

Deploying the NSG

As we mentioned earlier, an NSG allows us to filter inbound and outbound traffic. In this
section, we will deploy our NSG and look at the inbound and outbound rules that are
created by default:

1. Copy and paste *line 16*, `New-AzResourceGroupDeployment -Name`
    `"AdxChapter10NSGDeployment" -ResourceGroupName`
    `$resourceGroupName -TemplateFile ./templates/nsg.json`
    `-TemplateParameterFile ./parameters/nsg.params.json`, into
    Cloud Shell and hit *Enter*. The deployment may take a few minutes. Once it is
    complete, you will see the provisioning state, similar to the output you saw when
    creating the resource group.

2. Open a new browser tab, go to `https://portal.azure.com`, and click on the `adx-subnet-delegation-rg` resource group to see the resources. You should see the newly created NSG, as shown in the following screenshot:

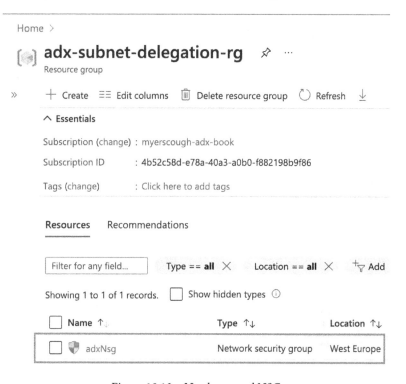

Figure 10.13 – Newly created NSG

3. Click the NSG. On the **Overview** blade, you should see a list of the inbound and outbound security rules, as shown in the following screenshot:

Priority ↑	Name ↑↓	Port ↑↓	Protocol ↑↓	Source ↑↓	Destination ↑↓	Action ↑↓
∨ Inbound Security Rules						
65000	AllowVnetInBound	Any	Any	VirtualNetwork	VirtualNetwork	✅ Allow
65001	AllowAzureLoadBalancer…	Any	Any	AzureLoadBalancer	Any	✅ Allow
65500	DenyAllInBound	Any	Any	Any	Any	❌ Deny
∨ Outbound Security Rules						
65000	AllowVnetOutBound	Any	Any	VirtualNetwork	VirtualNetwork	✅ Allow
65001	AllowInternetOutBound	Any	Any	Any	Internet	✅ Allow
65500	DenyAllOutBound	Any	Any	Any	Any	❌ Deny

Figure 10.14 – Default security rules

In the *Filtering traffic with NSGs* section, we will look at NSGs in a bit more detail and will learn how to add new security rules to filter traffic. But before we do that, let's deploy our route table.

Deploying the route table

The next resource we need to deploy is the route table. As we mentioned earlier, the route tables allow us to create user-defined routes. We will not be creating any user-defined routes. The routes will be created automatically when we deploy our ADX cluster but we must specify a route table. The route table is not automatically created for us. Let's get started:

1. Switch over to your Cloud Shell tab.

2. You will need to declare the `$resourceGroupName` and `$location` variables again if you are prompted to reconnect to your shell after being idle. You can do this by copying and pasting *lines 9* and *10* from `adx-arm-deploy.ps1` again into your Cloud Shell.

3. Copy and paste *line 17* from `adx-arm-deploy.ps1` – that is, `New-AzResourceGroupDeployment -Name "AdxChapter10RouteTableDeployment" -ResourceGroupName $resourceGroupName -TemplateFile ./templates/routeTable.json -TemplateParameterFile ./parameters/routeTable.params.json` – into Cloud Shell and hit *Enter*. The deployment may take a couple of minutes and will create a new route table called `adxRouteTable`.

We will look at the route table once we have deployed our ADX cluster. Next, let's deploy our virtual network.

Deploying the virtual network

The next resource we need to deploy is the virtual network and the subnet. The following steps demonstrate how to deploy a new virtual network and subnet and associate our NSG with the subnet using the ARM templates.

Copy and paste *line 20* from `adx-arm-deploy.ps1` – that is, `New-AzResourceGroupDeployment -Name "AdxChapter10VNETDeployment" -ResourceGroupName $resourceGroupName -TemplateFile ./templates/vnet.json -TemplateParameterFile ./parameters/vnet.params.json` – into Cloud Shell and hit *Enter*. The deployment will create a new virtual network called `adxVnet` and set our subnet, `adxSubnet`, as a delegated subnet for ADX clusters (`Microsoft.Kusto/clusters`).

You can view your virtual network in the Azure portal and verify that the subnet has been created and associated with our NSG and that the route table has been delegated, as shown in the following screenshot:

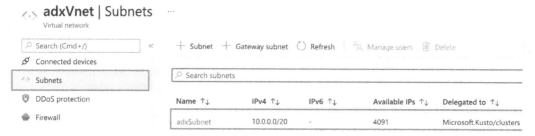

Figure 10.15 – Subnet delegated for ADX clusters

Now that we have deployed our virtual network, the next step is to deploy our public IP addresses. Once we have deployed the IPs, we can deploy our ADX cluster in the virtual network. In the next section, we will deploy one public IP address for the Kusto engine and one for ingestion.

Deploying the public IP addresses

The next set of resources we need are the public IP addresses. ADX requires two public IP addresses – one for the Kusto engine and one for the ingestion endpoint. When we create ADX clusters, two URLs are generated – one for the engine, such as `https://adxmyerscough.westeurope.kusto.windows.net`, and one for ingestion, such as `https://ingest-adxmyerscough.westeurope.kusto.windows.net`. Let's get started:

1. Copy and paste *line 23* from `adx-arm-deploy.ps1` – that is, `New-AzResourceGroupDeployment -Name "AdxChapter10EngineIpDeployment" -ResourceGroupName $resourceGroupName -TemplateFile ./templates/publicIp.json -TemplateParameterFile ./parameters/enginePublicIp.params.json` – into Cloud Shell and hit *Enter* to deploy the public IP address that will be used for the engine IP.

2. Copy and paste *line 24* from `adx-arm-deploy.ps1` – that is, `New-AzResourceGroupDeployment -Name "AdxChapter10IngestionIpDeployment" -ResourceGroupName $resourceGroupName -TemplateFile ./templates/publicIp.json -TemplateParameterFile ./parameters/dmPublicIp.params.json` – to deploy the public IP for ingestion. This IP address will be used for our data ingestion endpoint; for example, `https://ingest-adxmyerscough.westeurope.kusto.windows.net`.

Now that we have deployed our public IP addresses, the final step is to deploy our ADX cluster. In the next section, we will complete the deployment by deploying our ADX cluster.

Deploying the ADX cluster

The final step of the deployment is to deploy our ADX cluster into our virtual network. The following steps explain how to deploy an ADX cluster in our virtual network and how to examine the updates the deployment makes to our NSG and route table.

Copy and paste *line 27* from `adx-arm-deploy.ps1` – that is, `New-AzResourceGroupDeployment -Name "AdxChapter10ClusterDeployment" -ResourceGroupName $resourceGroupName -TemplateFile ./templates/adxCluster.json -TemplateParameterFile ./parameters/adxCluster.params.json` – into Cloud Shell and hit *Enter*. The deployment can take up to 15 minutes.

Once the deployment is complete, close Cloud Shell and open the `adx-subnet-delegation-rg` resource group in the Azure portal. We can visualize the deployment by clicking **Resource visualizer**, from the resource group's properties panel, as shown in the following screenshot:

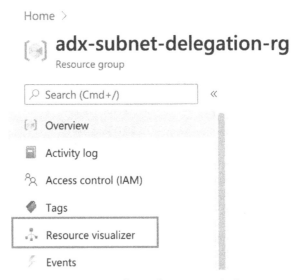

Figure 10.16 – Visualizing the resource deployment

The resource visualizer tool generates a diagram showing the dependencies between the resources that have been deployed in the resource group:

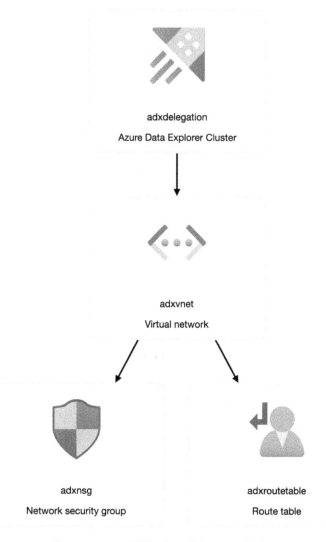

Figure 10.17 – ADX subnet delegation deployment

You may have noticed that the resource group contains extra resources, which we did not explicitly deploy. You should see four load balancers, as shown in the following screenshot:

Name ↑	Type ↑↓	Location ↑↓
adxDataIngestionPublicPIP	Public IP address	West Europe
adxDelegation	Azure Data Explorer Cluster	West Europe
adxEnginePublicPIP	Public IP address	West Europe
adxNsg	Network security group	West Europe
adxRouteTable	Route table	West Europe
adxVnet	Virtual network	West Europe
kucompute-adxdelegation-elb	Load balancer	West Europe
kucompute-adxdelegation-ilb	Load balancer	West Europe
kudatamgmt-adxdelegation-elb	Load balancer	West Europe
kudatamgmt-adxdelegation-ilb	Load balancer	West Europe

Figure 10.18 – The ADX cluster's load balancers

As you may recall from *Chapter 1, Introducing Azure Data Explorer*, an ADX cluster consists of two key components – the data management component and the engine – as shown in the following diagram:

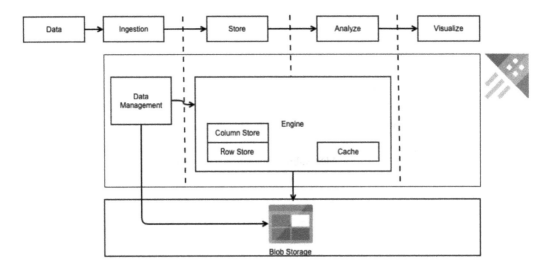

Figure 10.19 – ADX architecture

As we know, the data management and engine components are clusters of virtual machines and are load-balanced. Since we have delegated our cluster into our virtual network, the ADX cluster deployment deployed the cluster's load balancers into our subnet.

Now that the deployment is complete, let's try connecting to our ADX cluster:

1. Go to `https://dataexplorer.azure.com`.

2. Click **Add Cluster** and enter the ADX cluster's URI; for example, `https://adxdelegation.westeurope.kusto.windows.net`.

3. You will find that Data Explorer does nothing and appears unresponsive until the connection attempt times out:

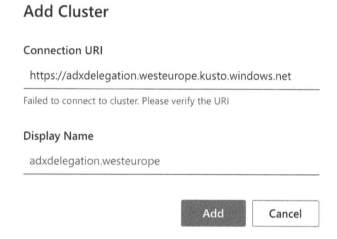

Figure 10.20 – Attempting to connect to the ADX cluster

The connection attempt failed because we have deployed the ADX cluster to our virtual network and we need to explicitly allow traffic from our IP address to connect. In the next section, we will look at the security rules that were added to our NSG as part of the ADX cluster deployment and learn how to enable traffic from our public IP address.

Filtering traffic with NSGs

When we deployed the ADX cluster, the deployment also updated our route table and NSG. The route table was updated to route specific cluster traffic to the internet, as shown in the following screenshot:

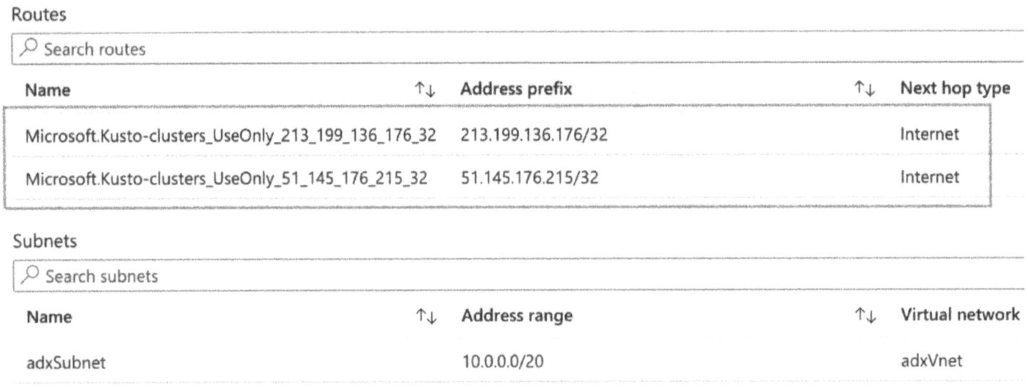

Routes

Name	↑↓	Address prefix		↑↓	Next hop type
Microsoft.Kusto-clusters_UseOnly_213_199_136_176_32		213.199.136.176/32			Internet
Microsoft.Kusto-clusters_UseOnly_51_145_176_215_32		51.145.176.215/32			Internet

Subnets

Name	↑↓	Address range		↑↓	Virtual network
adxSubnet		10.0.0.0/20			adxVnet

Figure 10.21 – User-defined routing

These routes ensure our cluster can still communicate with Azure Monitor and ADX cluster management.

In the next section, we will learn what they are, how they work, and why we should use them. Once we understand the theory, we will update our NSG by adding a new security rule to allow traffic to our ADX cluster.

Introducing NSGs

Before we add any security rules to our NSG, let's spend some time discussing what they are, how they work, and why we should use them.

NSGs are one of the fundamental security-related building blocks in Azure. NSGs allow us to filter both inbound and outbound traffic in an Azure Virtual Network. NSGs can be assigned to **subnets** and **Network Interface Cards** (**NICs**) if you are using virtual machines. NSGs consist of inbound and outbound security rules and each rule/tuple consists of the following seven attributes:

- **Priority**: NSG rules are evaluated from top to bottom. As soon as there is a match, the NSG stops checking the rules. Rules are sorted based on a priority, which is a number from 1 to 4,096, and the rules are sorted in ascending order. This means that rules with a lower priority number are higher in priority since they are evaluated first.

- **Name**: The name of the security rule.

- **Port**: The port or range of ports that should be evaluated.

- **Protocol**: The communication protocol. For example, you can filter TCP, UDP, and ICMP traffic, and so on.

- **Source**: The IP or IP range of the source/requester.

- **Destination**: The IP or IP range of the target destination.

- **Action**: Defines whether the traffic should be accepted or denied, as shown previously in *Figure 10.14*.

Both the inbound and outbound security rules are traversed from top to bottom until there is a matching rule. Once a security rule has been matched, the traffic is either allowed or denied, depending on the action in the security rule.

The source and destination values can also use service tags instead of IP addresses. **Service tags** represent different types of Azure resources, such as load balancers, and save you from having to list all the IP addresses for an Azure service. Trust me – there are many and they are updated regularly.

Every NSG has the following default inbound rules. These cannot be deleted but they can be overwritten by creating security rules with a lower priority number:

- AllowVNetInbound: Allows all inbound traffic on any port between virtual networks. This rule uses the VirtualNetwork service tag to represent the source and destination addresses.

- AllowAzureLoadBalancerInbound: Allows incoming traffic from Azure load balancers.

- DenyAllInbound: This rule denies all other inbound access.

All NSGs have the following outbound rules, which, like the default inbound rules, cannot be deleted:

- AllowVnetOutBound: Allows all outbound traffic on any port between virtual networks.

- AllowInternetOutBound: Allows all outbound internet traffic from the virtual network.

- DenyAllOutbound: This rule denies all other outbound traffic.

When deploying ADX clusters into subnets, the NSG associated with the subnet is updated with several inbound and outbound security rules. This allows the cluster's nodes to communicate with one another and communicate with other Azure resources, such as Azure Monitor and Azure Storage.

If we look closely at the inbound rules, we will see that there are no rules that allow traffic from the internet. The `DenyAllInBound` rule ensures that any traffic originating from any source other than Azure load balancers, virtual networks, and two IP address ranges is blocked.

In the next section, we will learn how to create an inbound security rule to allow traffic from our public IP so that we cannot connect to our ADX cluster via the internet.

Creating inbound security rules

The following steps will demonstrate how to create an inbound security rule that will allow inbound traffic from your public IP address:

1. Go to `https://portal.azure.com`.

2. Click on your NSG; for example, **adxNSG**.

3. In the NSG's properties pane, under **Settings**, click **Inbound security rules**, as shown in the following screenshot:

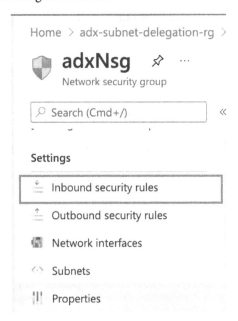

Figure 10.22 – Configuring inbound security rules

4. Click **+ Add**.

5. Set **IP Addresses** to **Source** from the drop-down box.

6. Enter your public IP under **Source IP addresses/CIDR ranges**. The easiest way to determine your public IP is to google `what's my IP?`.

7. Keep **Source port ranges** set to `*`. The asterisk denotes any port.

8. Keep **Destination** set to **Any** and **Service** set to **Custom**.

9. Set **Destination port ranges** to `*`.

10. Keep **Protocol** set to **Any**.

11. As we mentioned earlier, NSG security rules are sorted based on their priority in ascending order, and each rule is checked until a match is found. The default NSG rules start at 65,000 and the rules that are added by ADX range from 100 to 104. We want our rule to be somewhere in-between the ADX rules (100 – 104) and the NSG default rules (65,000 – 65,500). For example, set the value of **Priority** to `114`.

12. Give your rule a name, such as `JasonsLaptop`, and optionally add a description.

13. Click **Add** to add the rule to the NSG.

We can verify that we successfully created an inbound security rule by trying to connect to our ADX cluster using the ADX Web UI:

1. Go to `https://dataexplorer.azure.com`.

2. Click **Add Cluster** and enter your cluster's URL; for example, `https://adxdelegation.westeurope.kusto.windows.net`.

As shown in the following screenshot, I can connect to my ADX cluster successfully, which means I am ready to create my database tables since the new NSG rule allows traffic from my public IP address:

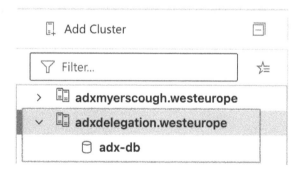

Figure 10.23 – Connected to our ADX cluster

In this section, we learned how to filter network traffic destined for our ADX cluster by creating new inbound security rules for our NSG to allow traffic from our laptops/ workstations.

Summary

In this chapter, we learned about the importance of security, especially on the public cloud, and introduced some of the basic terminology and concepts you should be familiar with, such as the principle of least privilege and RBAC. Next, we introduced the concepts of identity management with AAD, explained the differences between security principals, users, groups, and service principals, and the different levels of access, such as the management plane versus the data plane.

Next, we demonstrated how to restrict access to your ADX cluster using virtual networks and subnet delegation.

Finally, we introduced NSGs and explained how to use them to filter inbound traffic. Then, we demonstrated how to route inbound traffic from your public IP.

In the next chapter, you will discover how to troubleshoot performance issues with queries and learn about the best practices for writing queries and managing your clusters.

Questions

Before moving on to the next chapter, test your knowledge by trying out these exercises. The answers can be found at the back of this book:

1. Assign the `contributor` role to one of your AAD users in the management plane.
2. Assign the `Database ingestor` role to a user.
3. What happens when you log in to the ADX Web UI and try to query the database as that user?

11
Performance Tuning in Azure Data Explorer

Azure Data Explorer (**ADX**) is designed for high performance without the need for performance maintenance activities. However, it can still experience slow performance when overwhelmed by the workload. Therefore, it is important to understand **performance tuning** to ensure we maintain the high performance we know ADX delivers. In the examples we have seen so far, we have not had to worry about performance. Our datasets have been relatively small and even with the larger datasets we used on the help cluster, performance has not been an issue. With that said, as you make your cluster available to end users so that they can run queries and generate reports, their usage patterns and the queries they write can collectively impact performance.

In this chapter, we will begin by introducing performance tuning. Then, we will introduce **workload groups**, learn how they work, and how they can help preserve cluster performance. We will also create a workload group to limit requests for a particular group of users.

Then, we will learn about **policy management** and discover how to configure and manage policies using KQL management commands to improve performance. For instance, we will learn how to manage the caching and retention policies.

Finally, we will discuss some best practices that you should take into consideration when developing your KQL queries.

In this chapter, we are going to cover the following main topics:

- Introducing performance tuning
- Introducing workload groups
- Introducing policy management
- Monitoring queries
- KQL best practices

Technical requirements

The code examples for this chapter can be found in the `Chapter11` folder of this book's GitHub repository: `https://github.com/PacktPublishing/Scalable-Data-Analytics-with-Azure-Data-Explorer.git`.

Introducing performance tuning

Before we jump into workload groups, let's spend a few moments thinking about *performance tuning* in general. In general, performance should not be an issue, given that ADX has been designed and optimized to be a big data service that is highly scalable and fast. As you ingest more and more data and allow more users and applications to query your clusters, you may experience some performance degradation. Therefore, it is important to beware of performance tuning concepts and what features ADX provides to help tune performance when the time comes.

Like troubleshooting, which we discussed in *Chapter 9, Monitoring and Troubleshooting Azure Data Explorer*, performance tuning can be considered as a process. The goal of performance tuning is to identify bottlenecks, troubleshoot their causes, and apply the features that are available to us, such as workload groups, cache policies, and so on, to eliminate bottlenecks. It is also important to understand there is a performance versus cost trade-off. Yes, we could use the highest ADX SKU, but there is a good chance you will have budget constraints.

Once you allow end users and applications to access your ADX cluster, you will want to have the ability to control and monitor their usage. We need to do this to prevent the users and applications from consuming all the resources, which can result in performance problems for other users.

In the next section, we will discover how to use workload groups to restrict requests to our clusters to help reduce the risk of performance issues.

Introducing workload groups

I remember working on a big data project where we had a wide range of end users and applications using our clusters. At one end of the spectrum, we had engineers executing ad hoc queries to analyze application logs, while at the other end, we had product management and customer support teams running complex reports by using integrations into third-party tools, such as **Power BI**, to gain insights into usage patterns and statistics. At the end of each month, the team would start to receive phone calls and tickets related to query and job performance. Users were complaining that their jobs were either not running or timing out. It turned out that the customer support team was running jobs and reports to generate billing information and that these jobs were resource-intensive and would consume all the resources, causing other jobs to be queued or time out. The only way to resolve the issue was to log into the cluster and kill the long-running tasks.

Managing and monitoring resource usage can be difficult. You may require third-party tools and you may have no means of controlling how end users use your system. The good news is that ADX simplifies managing and monitoring usage by using **workload groups**. Workload groups allow us to group queries and commands based on a classification criterion. Then, we can assign policies to those groups to control resource consumption, such as the request rate and resource limiting. For instance, we could create a policy to restrict the number of concurrent connections. Later in this chapter, we will demonstrate how to restrict the number of requests a group of users can send per hour.

In the next section, we will learn about the built-in workload groups and how workload groups work.

How workload groups work

Workload groups allow us to group requests/queries and allow us to define constraints such as request limits, connection limits, and resource limits via workload group policies. This allows us to control access, monitor performance, and prevent users from consuming all our cluster's resources.

As shown in *Figure 11.1*, the end user sends a request – typically a **Kusto Query Language** (**KQL**) query – using one of the many clients, such as the Data Explorer Web UI and Azure portal. The requests are then processed by the **Request Classification Policy** (**RCP**), which is responsible for assigning the client's request to one of the workload groups. ADX allows us to define up to 10 custom workload groups and has a default and internal workload group already defined, which we will discuss shortly. If there are no custom workload groups, the requests are assigned to the default workload group.

Each workload group has an assigned workload group policy, which defines the constraints that will be enforced during execution. Policies are configurations for the subcomponents of our ADX cluster. For instance, you could define a policy to limit the number of requests coming from Power BI or you could define a policy to restrict the query execution time for a group of users.

It is important to understand how policies allow us to configure specific components of our clusters:

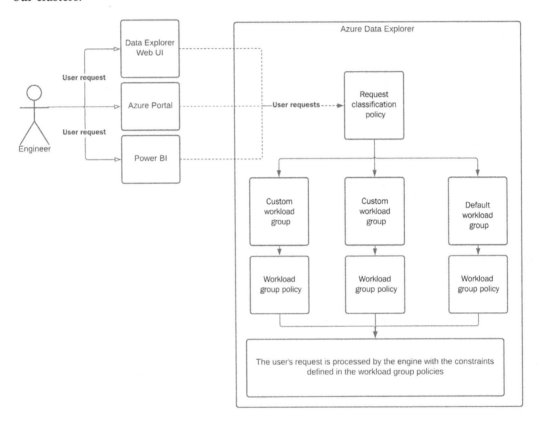

Figure 11.1 – Workload group workflow

Now, we will define the terms and discuss the components required to configure workload groups.

Request Classification Policy: This is responsible for processing all incoming requests and assigning them to the appropriate workload groups. If no request classification policy is defined, all the requests are assigned to the default workload group. The request classification policy has two important properties we need to set when creating our policy. The first is a Boolean called `IsEnabled`, which is used to enable the policy, while the second is `ClassificationFunction`, which is the body of our function and contains the logic for assigning requests to the different workload groups. `ClassificationFunction` is where we define our logic for assigning the incoming request/queries to the correct workload groups. `ClassificationFunction` has access to a property bag called `request_properties` that we can use to obtain more information regarding the incoming request. A property bag is a collection of key/value pairs. The following table summarizes the properties that are available to us:

Name	Description
`current_ database`	The name of the database that the end user wants to interact with.
`current_ application`	The name of the application that sent the request.
`current_ principal`	The name of the security principal that sent the request.
`query_ consistency`	This property can be used when you are using one of the SDKs, such as C#, Java, or Python.
`request_ description`	This is the custom text the author of the request can include. This is something that can be done when you use one of the SDKs, such as C#, Java, or Python.
`request_text`	The obfuscated text of the request.
`request_type`	The type of request. This can either be a query or a command.

Table 11.1 – request_properties properties

It is important to remember that the request classification policy is evaluated for every request that is sent to ADX, so please ensure that `ClassificationFunction` is as lightweight as possible; otherwise, you could impact performance.

Workload Groups: These are logical containers that allow us to groups queries. We can apply constraints to these groups via workload policies. ADX allows us to create up to 10 custom workload groups and comes with two built-in groups: **default** and **internal**.

Default: By default, all incoming requests are assigned to the default group. ADX allows us to modify the group assignment criteria and the workload group policies associated with this group.

Internal: As the name suggests, the internal group is used for internal use only by the ADX clusters. This group is completely managed by the cluster, and we are unable to make any changes to the group assignment criteria or the workload group policies associated with this group.

Workload Policies: These are the policies that we define and assign to the workload groups to ensure we have granular control over the incoming requests. In the next section, we will discover how to create workload groups and create a workload policy to restrict the number of requests per hour for a specific **Azure Active Directory (AAD)** group.

Creating custom workload groups

In this section, we will learn how to create workload groups and create a policy to restrict the number of requests per hour to a specific AAD group. In *Chapter 8, Data Visualization with Azure Data Explorer and Power BI*, we learned how to create AAD users. The two users I created, Harrison and James, will be used in this example. If you skipped the user account creation process, please return to *Chapter 8, Data Visualization with Azure Data Explorer and Power BI*, and create two user accounts.

Once you have the user accounts, we will create two new AAD groups— one called `PremiumUsers` and another called `TrialUsers`. We will use these groups and restrict the number of requests per hour for users that are members of `TrialUsers`.

> **Note**
>
> Please note that you must be authorized with `AllDatabasesAdmin` permissions to manage workload groups. For more information regarding authorization, please see *Chapter 10, Azure Data Explorer Security*.

In the next section, we will create the `PremiumUsers` and `TrialUsers` groups in the Azure portal.

Creating AAD groups

In this section, we will create our AAD groups via the Azure portal and add the users that you created in *Chapter 8, Data Visualization with Azure Data Explorer and Power BI*, to the groups. Let's get started:

1. Log in to the Azure portal (`https://portal.azure.com`).
2. Click on the hamburger icon on the top-left corner of the page. Then, click on **All services**, search for `active directory`, and click **Azure Active Directory**.

3. In the **Properties** panel, under **Manage**, click **Groups** and then click **+ New group**.

4. Leave **Group type** set to **Security** and enter a name for your first group; for example, `PremiumUsers`.

5. **Group description** is optional, so go ahead and click **Create** to complete the creation process.

6. Repeat *Steps 3* to *5* and create a new group called `TrialUsers`.

In this section, we created our groups. In the next section, we will add our AAD users to our groups.

Assigning AAD users to groups

Now that we have our AAD groups, the next step is to assign the users to our groups. In *Chapter 8, Data Visualization with Azure Data Explorer and Power BI*, I created two users called Harrison and James. In this example, we will add Harrison to `PremiumUsers` and James to `TrialUsers`:

1. Log in to the Azure portal (`https://portal.azure.com`).

2. Click **All services**, search for `active directory`, and click **Azure Active Directory**.

3. In the properties panel, under **Manage**, click **Users** and click the user you want to add to the `PremiumUsers` group. For example, I will add `Harrison`.

4. In the properties panel, under **Manage**, click **Groups** and click **+ Add memberships**.

5. Search for the `PremiumUsers` group, click the group, and then click **Select** to complete the assignment.

6. Repeat *Steps 3* to *5* and add your second user to the `TrialUsers` group. For example, I added `James` to `TrialUsers`.

In this section, we added our users to the AAD groups. In the next section, we will create our workload group and the policy for restricting the number of requests.

Creating workload groups

In this section, we will create a workload group with an initial policy that will restrict the number of current requests and the number of requests per hour:

1. Log in to the Data Explorer web UI (`https://dataexplorer.azure.com`).

2. Click **File**, then **Open**. From here, open `Chapter11/workloads/workloadGroup.kql` and execute the query by clicking **Run**.

The output of the query is shown in the following screenshot:

Figure 11.2 – Custom workload group creation

Lines 4–11 in the preceding screenshot define the limits for the maximum number of concurrent sessions. In this case, we set the limit to 5. *Lines 12–20* define the maximum number of requests per hour.

In the next section, we will learn how to create a request classification policy and assign incoming requests to our workload group.

Creating request classification policies

In this section, we will create a new request classification policy that will assign requests from users of `TrialUsers` to our `Packt Demo` workload group. These requests will be constrained to *one request per hour*. The following KQL management command enables our custom request classification policy:

```
.alter cluster policy request_classification
'{"IsEnabled":true}' <|
    iff(current_principal_is_member_
of('aadgroup=TrialUsers;27447925-1f0e-41b6-b01f-973eaab478b0'),
"Packt Demo","default")
```

The `iff(current_principal_is_member_of('aadgroup=TrialUsers;27447925-1f0e-41b6-b01f-973eaab478b0'), "Packt Demo","default")` command of `ClassificationFunction` checks if the requestor is a member of the `TrialUsers` group. The GUID after the semi-colon is our tenant ID. If the user is a member of `TrialUsers`, the request is assigned to the `Packt Demo` workload group; otherwise, the request is assigned to the default group.

> **Note**
>
> **GUIDs**, also known as **globally unique identifiers**, are popular identifiers that are used in Microsoft/Windows development. The IDs are made up of time, your network MAC address, which is also a unique ID, and your CPU information, such as clock speed. If you are interested in learning how GUIDs are generated, I recommend looking at the original RFC document, *RFC 4122*: `https://www.ietf.org/rfc/rfc4122.txt`.

In the next section, we will log into the Data Explorer Web UI as James or whichever user you assigned to the `TrialUsers` group and try executing more than one query.

Verifying workload groups

Before we log in as James, let's verify that our request classification policy has been enabled on our cluster. The following KQL management query returns the current request classification policy's configuration:

```
.show cluster policy request_classification
```

As shown in the following screenshot, on *lines 5* and *6*, our policy is enabled and our `ClassificationFunction` has been implemented:

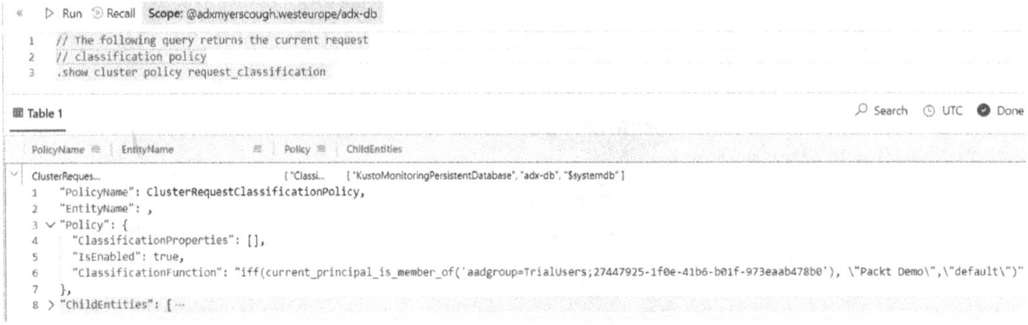

Figure 11.3 – Verifying the request classification policy's configuration

Now that we have verified that our request classification policy is enabled, the next step is to verify that it is working. When we log in as James, we should be able to execute one query and we should receive an error message if we attempt to run it more than one.

The following steps have you log in as James. You should log in as the user you assigned to your `TrialUsers` group and verify if our policy is working. Let's get started:

1. Log in to `https://dataexplorer.azure.com` using the user account that is a member of the `TrialUsers` group; for example, James (`james@myerscough.dev`).

2. Connect to our ADX cluster; for example, `https://adxmyerscough.westeurope.kusto.windows.net/`:

    ```
    EnglishPremierLeague
    | take 100
    ```

3. Try executing the query for a second time. As shown in the following screenshot, you should receive an error message, informing you that you have exceeded your request count quota:

Figure 11.4 – Workload group constraints

Here, we can see that our workload group and policies have been configured correctly. We can also verify James' subsequent queries were throttled by logging into the Data Explorer Web UI and executing the following query:

```
.show commands-and-queries
| where WorkloadGroup == "Packt Demo"
```

As shown in the following screenshot, the query returns a list of queries that have been executed against our cluster. We can see that the first query was completed successfully and that the subsequent queries were throttled:

« ▷ Run ⟳ Recall **Scope:** @adxmyerscough.westeurope/adx-db

```
8    .show commands-and-queries
9    | where WorkloadGroup == "Packt Demo"
```

⊞ Table 1   🔍 Search ⏱

	ClientActivityId	Comman...	Text	Database	State	FailureReason
>	KustoWebV2:ba3e3445...	Query	EnglishPremierLeague	take 100	adx-db	Completed	[none]
>	KustoWebV2:b192e73a...	Query	EnglishPremierLeague	take 100	adx-db	Throttled	The request was denied due to exceeding quota limitations. ...
>	KustoWebV2:8cfc41c7-...	Query	EnglishPremierLeague	take 100	adx-db	Throttled	The request was denied due to exceeding quota limitations. ...
>	KustoWebV2;36af7627-...	Query	EnglishPremierLeague	take 100	adx-db	Throttled	The request was denied due to exceeding quota limitations. ...
>	KustoWebV2;22533fa9-...	Query	EnglishPremierLeague	take 100	adx-db	Throttled	The request was denied due to exceeding quota limitations. ...

Figure 11.5 – Verifying that the requests were throttled

Before we wrap up this section, it is important to note that we can disable and re-enable the request classification policy at any time by toggling the `IsEnabled` variable. The following KQL management query disables our request classification policy and removes the request limit restrictions from the `TrialUsers` group:

```
.alter cluster policy request_classification
'{"IsEnabled":false}' <|
    iff(current_principal_is_member_
of('aadgroup=TrialUsers;27447925-1f0e-41b6-b01f-973eaab478b0'),
"Packt Demo","default")
```

In this section, we learned how to configure workload groups so that we can apply restrictions to groups of queries. This allows us to preserve the performance of our cluster and reduce the risk of bad performing queries disrupting our overall cluster performance.

In the next section, we will learn how to configure our caching and retention policy to help improve cluster performance.

Introducing policy management

ADX supports **performance tuning** via policies. These policies allow us to configure individual components of our cluster such as **caching**, **ingestion**, and **retention**. As you may recall from *Chapter 2*, *Building Your Azure Data Explorer Environment*, a lot of these settings can be set at the management plane level. The great benefit of policies is that we can configure these individual components without having to authorize contributor access at the management plane level. Administrators of the ADX databases, at the data plane level, can configure these policies.

In this section, we will demonstrate how to configure caching and retention policies using KQL management commands. For a complete list of policies that can be configured, please see the ADX documentation: `https://docs.microsoft.com/en-us/azure/data-explorer/kusto/management/`. The general syntax for managing policies is the same for all policies, so once you know how to configure one, configuring others should be straightforward.

Managing the cache policy

One of the reasons ADX is regarded as a high-performance big data solution is because of its hot cache. As we discussed in *Chapter 1*, *Introducing Azure Data Explorer*, ADX clusters use two types of storage – hot and cold. **Hot storage** is our high-speed cache that is attached directly to our engine nodes, while **cold storage** is an Azure storage account/data lake that is managed behind the scenes. To ensure high performance, ADX tries to ensure data is always loaded into the storage/the cache.

There are a couple of scenarios where you should tune the caching policy:

- The first is regarding *cost*. Hot storage is a lot more expensive than cold storage. In fact, at the time of writing, hot storage is 45 times more expensive. As we will see in the next chapter, *Chapter 12*, *Cost Management in Azure Data Explorer*, **Azure Advisor** is an Azure service that makes recommendations based on your Azure usage to help you optimize your deployments with regards to security, cost, and operational excellence. Under my subscription, Azure Advisor recommends that I reduce the caching window since it has not been used lately to help save on costs.

- The second is *performance*. You may need to expand the caching window, or you may want to cache data from a specific time window. For instance, we may want to analyze some logs from January 2021. We can update the caching policy to include data from January 2021 in the cache. We will demonstrate how to do this later in this chapter.

In the next section, we will demonstrate how to query the current caching policy that has been defined for databases and tables.

Viewing cache policies

To view the cache policies, we can use the `.show` KQL management command. For example, the following KQL management query displays the current caching policy for the `EnglishPremierLeague` table in our ADX cluster:

```
.show table EnglishPremierLeague policy caching
```

As shown in the following screenshot, on *line 3*, there is currently no policy set at the table scope:

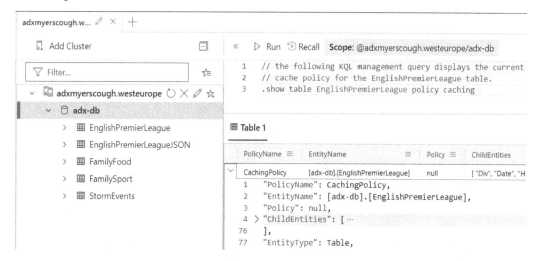

Figure 11.6 – Current caching policy for the EnglishPremierLeague table

By default, the tables inherit the policies of the databases. In this instance, if we check the policy at the database scope level, we can see that we have a policy that defines 30 days. The following KQL management query returns the caching policy for the `adx-db` databases in our ADX cluster:

```
.show database ['adx-db'] policy caching
```

The following screenshot shows the results of the previous query. As you can see, *lines 11* to *17* show a list of child entities that inherit the policy:

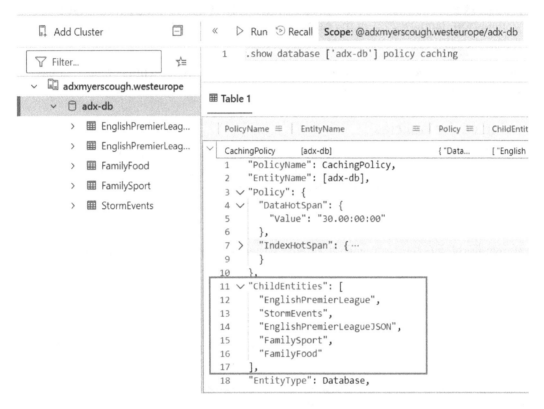

Figure 11.7 – ADX database caching policy

In the next section, we will demonstrate how to update the caching policy.

Modifying cache policies

Azure Advisor is recommending that I reduce the cache window to save on costs since I have not been querying the data. Rather than update the database caching policy, we will set a policy on the EnglishPremierLeague table. Policies set on the tables override the policies defined at the database scope level. We can use the .alter KQL management command to update the cache policy and assign the size of our cache to a variable named hot. The following query sets a cache policy on our EnglishPremierLeague table and sets the cache window to 5 days:

```
.alter table EnglishPremierLeague policy caching hot = 5d
```

As shown in the following screenshot, on *line 5*, the new cache policy has been set to 5 days (`5.00:00:00`):

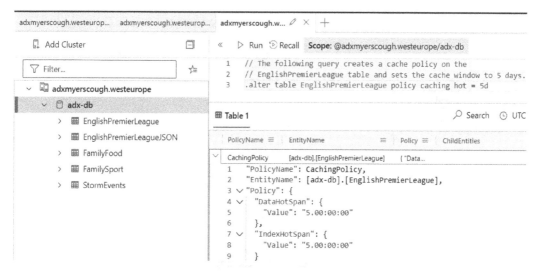

Figure 11.8 – The EnglishPremierLeague table's new cache policy

Data will be kept in the hot cache in the `EnglishPremierLeague` table for 5 days instead of 30 days. This does help you save on costs but you may experience performance issues if you are querying data outside of the 5 day window. For example, our cache is configured to 5 days. If we query for 7 days of data, then ADX will have to fetch data from the cool storage.

We can also delete policies by using the `.delete` KQL management command. The following query will remove the 5-day caching policy we just created for the `EnglishPremierLeague` table:

```
.delete table EnglishPremierLeague policy caching
```

As shown in the following screenshot, the caching policy has been removed and the policy value on *line 3* has been set to `null`:

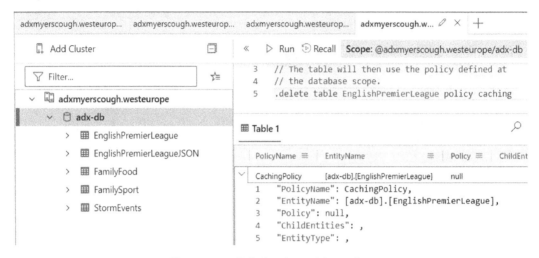

Figure 11.9 – Deleting the caching policy

The table will now use the policy that was defined at the database scope level. When defining caching policies, we can also specify data from a specific period to be loaded into the cache. For example, let's imagine we wanted to define the cache window as 30 days and we wanted to load data from June 2021. We can define specific periods by updating the `hot_window` variable. For instance, the following query creates a cache policy on our `EnglishPremierLeague` table and loads the data from June 2021:

```
.alter table EnglishPremierLeague policy caching
    hot = 30d,
    hot_window = datetime(2021-06-01) .. datetime(2021-06-30)
```

As shown in the following screenshot, the cache window has been set to 30 days and on *lines 10* to *13*, we have defined a time when data should be loaded into the cache:

Figure 11.10 – Loading specific data into the hot cache

This can be useful when you need to analyze logs files from the past that would be considered too old to be loaded into the cache. I have worked on issues in production where I needed fast access to log files from a few months earlier and have used such policies to improve the performance of my analysis.

In this section, we learned how to manage our caching policies via KQL management commands rather than via the management plane in the Azure portal. We also learned how to configure the hot_window variable to load data from specific periods into our cache. At the time of writing, this capability is not configurable in the Azure portal/management plane.

In the next section, we will discover how to manage our data retention periods at the data plane scope level via KQL management commands.

Managing retention policies

Another policy that is relevant to performance is the retention policy. The key difference between the **cache policy** and the **retention policy** is that the cache policy is concerned with how long data is stored in the hot cache. Data that falls outside the cache window is flushed out of the hot cache. On the other hand, the **retention policy** is concerned with how long our data is available in our ADX cluster. The obvious benefit of retention policies is cost. As you know, we pay for the amount of data we ingest and store. The more data we ingest, the more expensive our monthly costs will be if we do not use retention policies. As we discovered in *Chapter 1*, *Introducing Azure Data Explorer*, data is stored in shards/extents and the fewer shards we have means the less processing we have to do. We will discuss the use of shards in the *Prioritizing time filtering* section.

In the next section, we will demonstrate how to query the current retention policies that have been defined for our databases and tables.

Viewing retention policies

To view our retention policies, we can use the `.show` KQL management command. For example, the following KQL management query displays the current retention policy for the `EnglishPremierLeague` table in our ADX cluster:

```
.show table EnglishPremierLeague policy retention
```

As shown in the following screenshot, no retention policy has been explicitly assigned to the `EnglishPremierLeague` table and like the cache policies, the tables inherit the policies defined at the database scope level:

Figure 11.11 – The EnglishPremierLeague retention policy

The following query returns the retention policy that has been assigned to our
adx-db database:

```
.show database ['adx-db'] policy retention
```

As shown in the following screenshot, on *lines 3* to *5*, we have a policy defined that has
a retention period of 365 days and has soft delete enabled (Recoverability), which
means we have up to 14 days to recover our data. We will discuss Recoverability in
the next section:

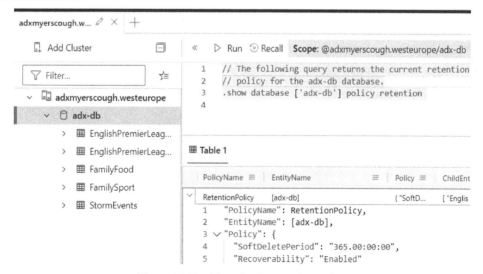

Figure 11.12 – The adx-db retention policy

In the next section, we will learn how to create and update policies and how we can apply the default policy.

Modifying retention policies

We can use the `.alter` KQL management command to create and update retention policies. The retention policy has the following attributes we can modify:

- `SoftDeletePeriod`: This defines the data retention period. The default value is 100 years. Shortly, we will discover how to enable the default policy.

- `Recoverability`: This sets the soft delete to enabled, which gives us up to 14 days to recover any data that has been deleted.

As we mentioned earlier, the default retention policy is 100 years with `Recoverability` enabled. While I do not recommend that you set your policy to the default in a real-world scenario due to the enormous costs that would be associated with such a retention period, the following query enables the default retention policy on our `EnglishPremierLeague` table:

```
.alter table EnglishPremierLeague policy retention "{}"
```

As shown in the following screenshot, on *lines 3 to 5*, we have added a retention policy to our `EnglishPremierLeague` table, and the retention period is set to 100 years (`36500` days):

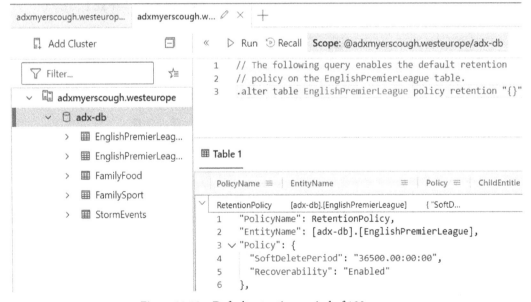

Figure 11.13 – Default retention period of 100 years

Since it is not practical to have such a retention policy due to the huge cost implications associated with retaining data for 100 years, let's go ahead and delete the policy using the .delete KQL management command. The following query demonstrates how to delete the retention policy from our EnglishPremierLeague table:

```
.delete table EnglishPremierLeague policy retention
```

As shown in the following screenshot, on *line 3*, the retention policy has been removed and the current policy has been assigned to null, which means that the table will inherit the retention policy defined at the database scope level:

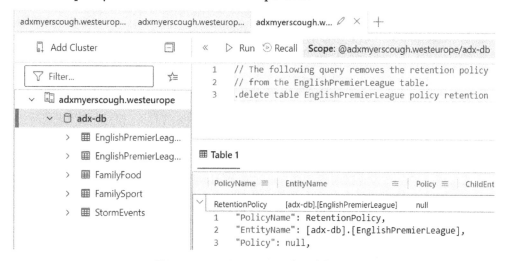

Figure 11.14 – Retention policy deletion

Before we move on to the next section, let's look at setting a retention policy of 90 days with Recoverability disabled and understand its impact:

```
.alter table EnglishPremierLeague policy retention
    "{\"SoftDeletePeriod\": \"90.00:00:00\",
\"Recoverability\": \"Disabled\"}"
```

As you can see, the syntax for setting retention policies is slightly different from the syntax for cache policies. For retention policies, we need to wrap our configuration in curly braces ({ }). As shown in the following screenshot, on *lines 4* and *5*, we have a retention policy configured to 90 days with `Recoverability` disabled, which means data cannot be recovered once it has been deleted after 90 days. So, be careful when using this setting:

Figure 11.15 – Creating a new retention policy of 90 days

In this section, we introduced retention policies, which can help with performance since it means we have less data to process. However, the primary benefit of retention policies is to save costs.

In the next section, we will learn how to monitor queries that have been executed by our ADX clusters.

Monitoring queries

At the beginning of this chapter, in the *Introducing performance tuning* section, we learned that performance tuning is a process and that one of the steps is to identify the root cause of performance bottlenecks. We also saw in *Chapter 9*, *Monitoring and Troubleshooting Azure Data Explorer*, that the ADX Insights page in the Azure Portal provides out-of-the-box dashboards for performance and other telemetry, which is are very useful.

We can use KQL management commands to view all the commands and queries that have been executed by our ADX cluster. These KQL management commands provide valuable insights into CPU consumption, query execution duration, the query/command being executed, and the state of the query's execution. The following query returns all the queries, sorted by execution duration, which have been executed on our ADX cluster:

```
.show queries
| project Text, Database, StartedOn, Duration, State,
```

```
FailureReason, TotalCpu, CacheStatistics.Disk.Misses,
WorkloadGroup, User, Principal
| sort by Duration desc
```

The following screenshot shows that one of the queries failed. Based on the `Text` column, we can see that someone tried to check for failed ingestions by executing `FailedIngestion | count`:

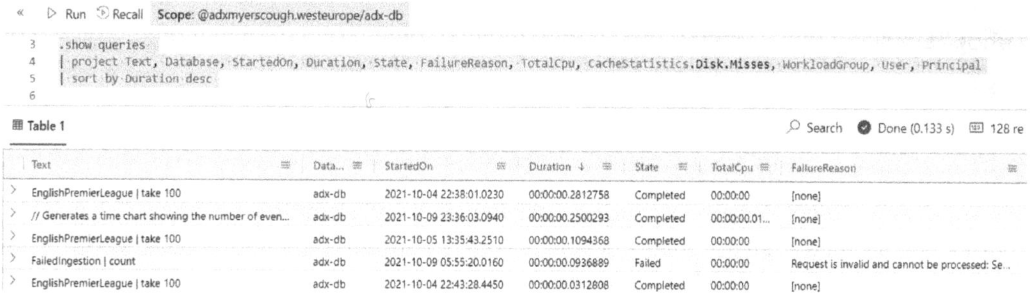

Figure 11.16 – Monitoring the query's execution

Using the knowledge we gained in *Chapter 5*, *Introducing the Kusto Query Language*, we can use the aggregation operators. For instance, the following command aggregates the `State` columns and returns a count of the different states:

```
.show queries
| summarize count() by State
```

As shown in the following screenshot, we can see that three states were returned – `Completed`, `Throttled`, and `Failed`:

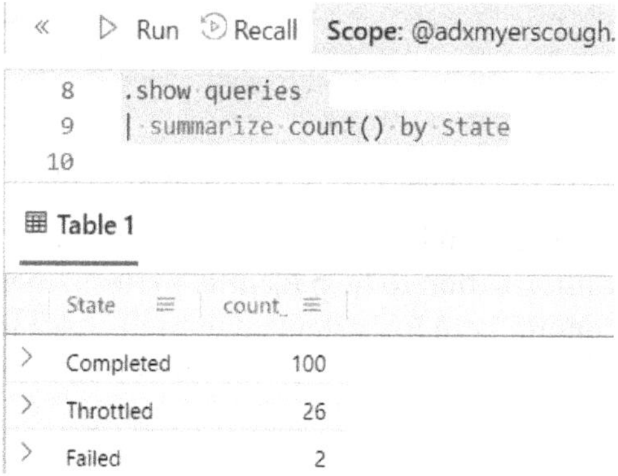

Figure 11.17 – Aggregating the query's execution state

The available information is returned by the KQL management commands. I would recommend building a dashboard to have an overview of query performance. One of the questions/exercises at the end of this chapter is to build such a dashboard.

The `.show queries` KQL management command returns information regarding queries. We can also return command execution information. For example, the following KQL management command returns the top 10 commands that have been executed on our cluster:

```
.show commands
| project Text, StartedOn, Duration, User, TotalCpu,
WorkloadGroup
| take 10
```

The results returned by `.show commands` are the internal commands that are executed by our cluster. Here, we can see that they are internal by checking the `WorkloadGroup` column, as shown in the following screenshot:

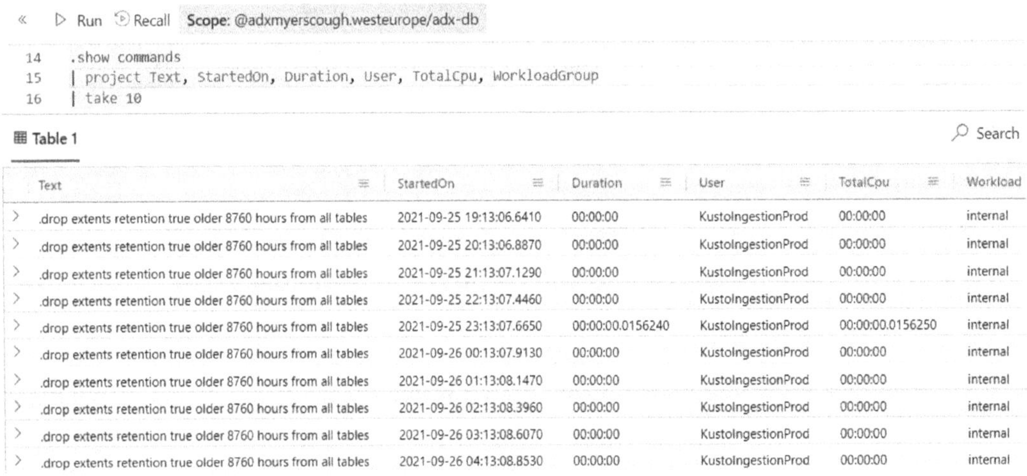

```
«   ▷ Run  ⏵ Recall  Scope: @adxmyerscough.westeurope/adx-db

14   .show commands
15   | project Text, StartedOn, Duration, User, TotalCpu, WorkloadGroup
16   | take 10
```

Text	StartedOn	Duration	User	TotalCpu	Workload
.drop extents retention true older 8760 hours from all tables	2021-09-25 19:13:06.6410	00:00:00	KustoIngestionProd	00:00:00	internal
.drop extents retention true older 8760 hours from all tables	2021-09-25 20:13:06.8870	00:00:00	KustoIngestionProd	00:00:00	internal
.drop extents retention true older 8760 hours from all tables	2021-09-25 21:13:07.1290	00:00:00	KustoIngestionProd	00:00:00	internal
.drop extents retention true older 8760 hours from all tables	2021-09-25 22:13:07.4460	00:00:00	KustoIngestionProd	00:00:00	internal
.drop extents retention true older 8760 hours from all tables	2021-09-25 23:13:07.6650	00:00:00.0156240	KustoIngestionProd	00:00:00.0156250	internal
.drop extents retention true older 8760 hours from all tables	2021-09-26 00:13:07.9130	00:00:00	KustoIngestionProd	00:00:00	internal
.drop extents retention true older 8760 hours from all tables	2021-09-26 01:13:08.1470	00:00:00	KustoIngestionProd	00:00:00	internal
.drop extents retention true older 8760 hours from all tables	2021-09-26 02:13:08.3960	00:00:00	KustoIngestionProd	00:00:00	internal
.drop extents retention true older 8760 hours from all tables	2021-09-26 03:13:08.6070	00:00:00	KustoIngestionProd	00:00:00	internal
.drop extents retention true older 8760 hours from all tables	2021-09-26 04:13:08.8530	00:00:00	KustoIngestionProd	00:00:00	internal

Figure 11.18 – Monitoring the command's execution

In this section, we learned how to gain insights into query and command execution on our cluster. We can use this information to help identify performance issues. For example, the `CacheStatistics` column is an important column to check to identify issues with the cache. If the cache's miss counts are high, then it could be an indicator to adjust your cache policy.

In the next section, we will discuss some of the best practices/habits that I think you will find useful as you write and develop your KQL queries.

KQL best practices

In this section, we will discuss some of the best practices that you should take into consideration when developing your KQL queries.

Version controlling your queries

The first recommendation is to **version control** all your queries, any scripts, and so on, that you may be using. Version control is a way to track changes in your source code – in our case, KQL queries. Not only does version control help us keep track of all source code changes that have been made to your code, but it also helps us share code with colleagues and friends. All the code examples that accompany this book are version controlled using a popular version control tool called Git and are hosted on `github.com`.

At one point in their careers, I am pretty sure all developers have looked at their old code and felt like they had no idea what the code does. Since version control keeps track of all changes, we can easily search the change history to understand why certain changes have been made. Let's take a look at the following screenshot:

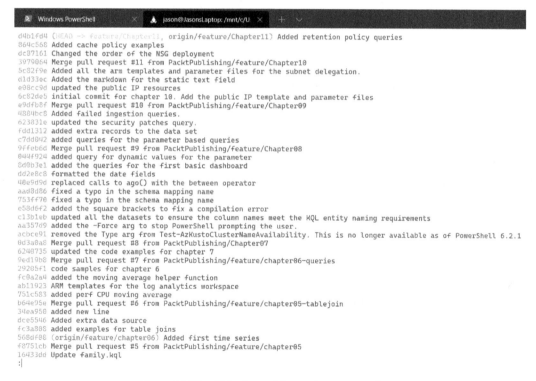

Figure 11.19 – Git's change history for our repository

The preceding screenshot displays the commit history for the repository that accompanies this book. You can view the commit history for the accompanying repository by executing the following `git` command:

```
git log --oneline
```

Prioritizing time filtering

If your datasets contain a date field, ensure your first `where` clause filters the time range in which you are interested. This will greatly improve your query's performance. The improvements are related to how ADX stores data. In *Chapter 1*, *Introducing Azure Data Explorer*, we learned that ADX stores our data in **shards**, also known as **extents**. By filtering our data on a specific time range, we exclude data outside of the time range, thus reducing the amount of data processing during execution time.

Best practices for string operators

The following are some of the best practices to consider when searching for and using string operators:

- Pay special attention to the string operators you use when searching data. For example, avoid using the `search` keyword and wildcards (`*`). The `search` keyword will search across all tables and rows, while the wildcard will search all columns. You will not experience performance issues when dealing with small datasets, but as your datasets grow, the impacts/latency will be more noticeable. Instead, be more specific! If you know what you are looking for, then explicitly specify the table and columns you want to search.

- Use case-sensitive operators where possible. Case-insensitive operators introduce overheard since the operator cannot simply compare two strings. It must prepare the strings to ignore the case. Remember that all the string operators have case-sensitive versions and have the `_cs` postfix. For example, the case-sensitive version of `has` is `has_cs`.

This section introduced some of the best practices to consider when writing your KQL queries. Now, let's summarize this chapter.

Summary

In this chapter, we started by introducing performance tuning and discovered that performance tuning is a process similar to troubleshooting. Next, we learned what workload groups are and how to configure them. We created an example workload group that restricted the number of requests that members from a specific AAD group can make. We discovered the three components that are required to use workload groups are the request classification policy, which is responsible for assigning requests to workload groups, the workload groups themselves, and the workload group policies that allow us to apply restrictions to requests, such as rate limiting.

Next, we discovered how to manage our cluster performance by managing the hot cache from the data plane. By managing the cache at the data plane, we can allow database administrators to tune performance, without giving them access to the management plane.

Next, we introduced the `.show queries` and `.show commands` KQL management commands. These management commands give us insights into query and command performance and can be used as an indicator that we may need to tune the components of our cluster, such as the hot cache.

Finally, we discussed some best practices that you should consider when developing your KQL queries. The most important were version control and time range filtering, which helps reduce the amount of data that is processed during query execution.

In the next and final chapter, we will discuss how we can manage our Azure costs.

Questions

Before moving on to the next chapter, test your knowledge by trying out these exercises. The answers can be found at the back of this book:

1. What is the purpose of workload groups?

2. Assuming that we have our request classification policy configured and enabled, what will happen when we execute the following query as a database admin?

```
.alter cluster policy request_classification
'{"IsEnabled":false}' <|
    iff(current_principal_is_member_
of('aadgroup=TrialUsers;27447925-1f0e-41b6-b01f-
973eaab478b0'), "Packt Demo","default")
```

3. Why should you filter your data based on a date field as early as possible in your query?

4. Create a dashboard in the Data Explorer Web UI and display the query execution metrics, such as the longest top 5 running queries, and aggregate the workload groups. Hint: use `.show queries` and review *Chapter 8, Data Visualization with Azure Data Explorer and Power BI.*

12
Cost Management in Azure Data Explorer

Azure provides access to a lot of powerful resources at the tips of our fingers—with a couple of clicks, you can have access to high-performing **virtual machines (VMs)**, **Structured Query Language (SQL)** servers, and, of course, **Azure Data Explorer (ADX)**. The problem is, if you do not pay close attention to your consumption, the cost can quickly grow, and you may end up with a huge bill at the end of the month. Luckily, Azure provides a couple of very useful features for managing your Azure costs.

In this chapter, we will begin by introducing **scaling** and how it relates to cost in Azure, then we will learn about the different ADX cluster **stock-keeping units (SKUs)** and how to select the correct SKU for your use case. We will then discuss ADX's scaling capabilities and what you should be aware of when using either manual or automatic scaling.

Next, we will introduce a useful Azure feature called **Azure Advisor**. It is, in my opinion, one of the least used features but can be quite useful in providing recommendations for your usage in Azure. For example, Azure Advisor provides recommendations for cost, security, reliability, and operational excellence.

Finally, we will introduce Azure's **cost management** features and learn how to view our invoices, generate billing reports and forecasts, and configure budgets and alerts so that we are notified when we are close to or have exceeded our budget.

In this chapter, we are going to cover the following main topics:

- Scaling and cost management

- Selecting the correct ADX cluster SKU

- Introducing Azure Advisor

- Introducing Cost Management + Billing

Technical requirements

There are no code examples in this chapter, but I would like to point out that pricing on Azure varies from region to region. All prices listed in the book are for the West Europe region and I have used my local currency, which is the **Euro** (**EUR**). Please note that at the time of writing, Azure Advisor only displays cost savings in **US Dollars** (**USD**).

Scaling and cost management

One of the aspects that we have not discussed in detail so far is **scaling**. One of the design principles of cloud computing is elasticity, and Azure allows us to scale our resources on demand. Scaling in the context of elasticity comes in two dimensions. One is **vertical scaling**, which refers to increasing the specification of a VM or ADX engine node. For instance, changing the engine SKU from `Standard_E64i_v3` to `Standard_E80ids_v4` is an example of vertical scaling.

The second dimension is horizontal scaling. **Horizontal scaling** refers to adding more VMs or engines. For instance, increasing the number of engines is a form of horizontal scaling.

ADX can take care of scaling for us and this is referred to as **auto-scaling**, but there can be up to 10 minutes of downtime. If downtime is an issue, then you can also use **manual scaling** and decide when to scale your cluster—for instance, you could manually scale your cluster outside of peak hours.

If you decide to use auto-scaling, please be aware that scaling can decrease and increase your Azure costs.

We have not demonstrated the use of auto-scaling in this book since the dev/test clusters do not support this. If you would like more information regarding auto-scaling, please see https://docs.microsoft.com/en-us/azure/data-explorer/manage-cluster-horizontal-scaling.

In the next section, *Selecting the correct ADX cluster SKU*, we will learn about the different SKU types that are available to us.

Selecting the correct ADX cluster SKU

We learned in *Chapter 1, Introducing Azure Data Explorer*, that ADX clusters consist of two VM clusters—the **engine cluster**, which is primarily responsible for querying our data, and the **data management cluster**, which is primarily responsible for data ingestion. When we build ADX clusters and specify the cluster size, we are specifying the VM SKU for our engine and data management clusters. The question is: *How do we select the right SKU?*

Azure provides several configurations, each optimized for different use cases. For instance, if you are ingesting huge volumes of data, then perhaps the storage-optimized SKUs are a better fit for your requirements. On the other hand, if you will be running a lot of jobs and queries, then one of the compute-optimized SKUs might be the right choice.

Before we explore the different SKUs, it is important to know Azure provides clusters with two different service levels. The first is dev/test, which is the cheapest option and provides no **service-level agreement** (**SLA**), thus Microsoft does not guarantee availability. The second is production, which is more expensive, but Microsoft provides an SLA of 99.9% availability. If Microsoft fails to meet the SLA, you are eligible for Azure credits as compensation.

In the next section, *Introducing dev/test clusters*, we will take a closer look at dev/test clusters, discuss when to use them, and examine their limitations.

Introducing dev/test clusters

Dev/test clusters, as their name implies are ideal for development and test environments. All the examples in this book have used the dev/test clusters primarily for cost reasons. As mentioned earlier, Microsoft does not provide an SLA for availability, so it is not recommended to use these clusters in your production environments. Dev/test clusters are the cheapest option available, but there are some limitations that you should be aware of.

One of the reasons Microsoft provides no SLA is due to the engine and data management components running as single VM instances. Another limitation is that horizontal scaling is not possible, which means you are unable to add more VMs to your engine cluster.

As shown herein *Table 12.1*, the dev/test clusters only support two VM SKUs, and the main difference is the size of the hot cache:

Name	Category	Cache	Central processing unit (CPU) cores	Random-access memory (RAM)	Minimum instance count	Maximum instance count	Monthly cost in EUR
Dev (No SLA) Standard_ D11_v2	compute-optimized	80 **gigabytes (GB)**	1	14 GB	1	1	€116.97
Dev (No SLA) Standard_ E2a_v4	compute-optimized	24 GB	1	16 GB	1	1	€93.57

Table 12.1 – Dev SKU configuration

While writing the book, I did not experience any problems with the dev/test cluster types. The clusters were always available, and I could successfully complete my development and testing without any issues.

If you need **high availability** (**HA**) or would like to use the other optimized VM SKUs such as storage and heavy compute, then you will have to use one of the production clusters, which costs more. In the next section, *Introducing production clusters*, we will take a look at the different VM SKUs that are available.

Introducing production clusters

In addition to the performance improvements and highly specialized VM SKUs, production clusters come with an SLA of 99.9% availability. However, the SLA comes at an extra cost since Microsoft requires at least two VMs for the engine and two VMs for the data management costs. Production clusters also support horizontal scaling, but the minimum number of VMs is two.

Production clusters support the following VM SKUs: compute-optimized, heavy compute, storage-optimized, and isolated compute.

Compute-optimized

Compute-optimized SKUs—or, rather, **D-series VMs**—are optimized for CPU-intensive operations such as high volumes of queries to process and have a balanced *CPU-to-memory ratio*. The Dv2 series runs second generation Intel® Xeon® Platinum 8272CL (Cascade Lake), Intel® Xeon® 8171M 2.1GHz (Skylake), Intel® Xeon® E5-2673 v4 2.3 GHz (Broadwell), or the Intel® Xeon® E5-2673 v3 2.4 GHz (Haswell).

The following table provides a list of all the available compute-optimized SKUs for the West Europe region. It is important to be familiar with the different types of SKUs that are available, both from a cost and performance perspective. As you can see, the biggest cost factor is the cache; therefore, it is important to monitor cache utilization factor:

Name	Category	Cache	CPU cores	RAM	Minimum instance count	Maximum instance count	Monthly cost in EUR
Standard_D11_v2	compute-optimized	80 GB	2	14 GB	2	8	€233.93
Standard_D12_v2	compute-optimized	160 GB	4	28 GB	2	16	€466.63
Standard_D13_v2	compute-optimized	317 GB	8	56 GB	2	1,000	€934.49
Standard_D14_v2	compute-optimized	628 GB	16	112 GB	2	1,000	€1,868.99

Table 12.2 – Compute-optimized SKU configuration

I recommend using the Dv2 series of machines if you or your end users will be running hundreds of queries on your cluster.

Heavy compute

As with the D-series, the E-series VMs are designed for computing and processing a lot of in-memory data. When you require fast query performance, then the E-series is a good choice given the size of the cache. The E-series VMs run on AMD EPYC processors. You can view more information about them here:

Name	Category	Cache	CPU cores	RAM	Minimum instance count	Maximum instance count	Monthly cost in EUR
Standard_E2a_v4	heavy compute	24 GB	2	16 GB	2	8	€467.86
Standard_E4a_v4	heavy compute	60 GB	4	32 GB	2	16	€374.29
Standard_E8a_v4	heavy compute	137 GB	8	64 GB	2	1,000	€748.58
Standard_E16a_v4	heavy compute	273 GB	16	128 GB	2	1,000	€1,497.16

Table 12.3 – Heavy compute-optimized SKU configuration

A difference between the D-series and the E-series is that the E-series supports both generation 1 and generation 2 VMs. If you are interested in learning more about the different VM generations, then I recommended looking at `https://docs.microsoft.com/en-us/windows-server/virtualization/hyper-v/plan/should-i-create-a-generation-1-or-2-virtual-machine-in-hyper-v`.

Storage-optimized

Storage optimized SKUs, as the name implies, are optimized for high disk throughput and **input/output (I/O)**. You should use this kind of SKU when you require your engine nodes to have large local storage volumes. As shown in *Table 12.4*, the local cache ranges from 1 **terabyte (TB)** to 4 TB. The L-series machines are designed specifically for running big data workloads, such as **SQL Server**, **Redis**, and **Cassandra**. The L-series also support local **Non-Volatile Memory Express (NVMe)** storage. For more information regarding the L-series VMs, please see `https://docs.microsoft.com/en-us/azure/virtual-machines/lsv2-series`:

Name	Category	Cache	CPU cores	RAM	Minimum instance count	Maximum instance count	Premium storage (PS) in TB	Monthly cost in EUR
Standard_DS13_v2 + 1 TB PS	storage-optimized	1 TB	8	56 GB	2	1,000	1	€1,185.26
Standard_DS13_v2 + 2 TB PS	storage-optimized	2 TB	8	56 GB	2	1,000	2	€1,436.02
Standard_DS14_v2 + 3 TB PS	storage-optimized	3 TB	16	112 GB	2	1,000	3	€2,621.28
Standard_DS14_v2 + 4 TB PS	storage-optimized	4 TB	16	112 GB	2	1,000	4	€2,872.04
Standard_E8as_v4 + 1 TB PS	storage-optimized	1 TB	8	64 GB	2	1,000	1	€999.34
Standard_E8as_v4 + 2 TB PS	storage-optimized	2 TB	8	64 GB	2	1,000	2	€1,250.11

Name	Category	Cache	CPU cores	RAM	Minimum instance count	Maximum instance count	Premium storage (PS) in TB	Monthly cost in EUR
Standard_ E16as_v4 + 3 TB PS	storage-optimized	3 TB	16	128 GB	2	1,000	3	€2,249.45
Standard_ E16as_v4 + 4 TB PS	storage-optimized	4 TB	16	128 GB	2	1,000	4	€2,500.22
Standard_ L4s	storage-optimized	650 GB	4	32 GB	2	16	0	€458.01
Standard_ L8s	storage-optimized	1.3 TB	8	64 GB	2	1,000	0	€916.03
Standard_ L16s	storage-optimized	2.6 TB	16	128 GB	2	1,000	0	€1,832.05
Standard_ L8s_v2	storage-optimized	1.7 TB	8	64 GB	2	1,000	0	€916.03
Standard_ L16s_v2	storage-optimized	3.5 TB	16	128 GB	2	1,000	0	€1,832.05

Table 12.4 – Storage-optimized SKU configuration

Isolated compute

As you know, Azure is a multi-tenant public cloud service, which means one physical host could be running VMs for multiple customers. In general, this is fine and does not cause problems. There are rare occasions where one customer's workload can impact other customers by consuming more of the physical host's resources than it should. This phenomenon is known as the **noisy neighbor** and is a common problem in multi-tenant environments. Azure provides **isolated compute**, whereby the VMs run on dedicated hardware and are not shared with multiple customers. If you need the guarantee of running on dedicated hardware, then this SKU is for you. You can view more details here:

Name	Category	Cache	CPU cores	RAM	Minimum instance count	Maximum instance count	Monthly cost in EUR
Standard_ E64i_v3	isolated compute	1.1 TB	64	432 GB	2	1,000	€5,387.81
Standard_ E80ids_v4	isolated compute	1.8 TB	80	504 GB	2	1,000	€8,520.03

Table 12.5 – Isolated compute SKU configuration

It should come as no surprise that the different SKUs are not priced the same, and the region you deploy to can also have an impact on the cost. For instance, the dev/test clusters in *West US 2 (€91.73)* are cheaper than in *West Europe (€116.97)*. Fortunately, Microsoft provides an **Azure pricing calculator** that you can use to generate cost estimates for your deployments. The calculator can be found at `https://aka.ms/ azurepricingcalculator`.

The following sequence of steps explains how to use the Azure pricing calculator:

1. Go to `https://aka.ms/azurepricingcalculator`.
2. Then, search for `Azure Data Explorer`, as shown in the following screenshot:

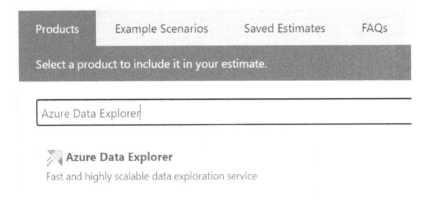

Figure 12.1 – Azure pricing calculator

3. Click **Azure Data Explorer**. A small notification will appear in the bottom-right corner of your screen, as shown in the following screenshot. Click **View** to be taken to the input form:

Figure 12.2 – Azure calculator notification

4. As shown in the following screenshot, based on your data requirements, the calculator tries to select the best VM SKU for your workloads. If you want to select the VM SKU manually, you can toggle the **AUTO-SELECT ENGINE INSTANCES** switch. Please keep in mind that you can only change the engine VM SKU. The data management SKU is selected automatically and cannot be changed:

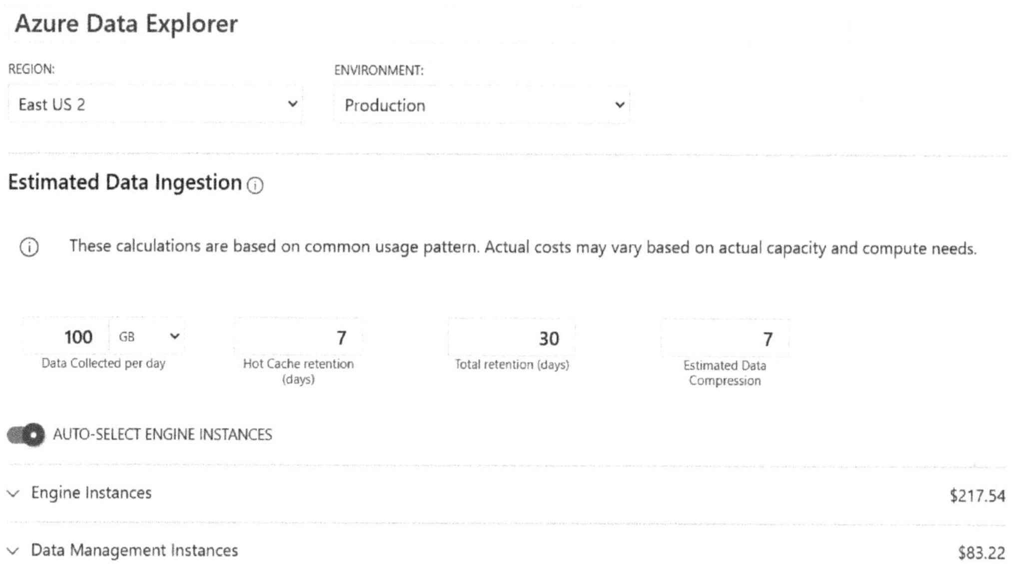

Azure Data Explorer

REGION:

East US 2   ⌄

ENVIRONMENT:

Production   ⌄

Estimated Data Ingestion ⓘ

ⓘ   These calculations are based on common usage pattern. Actual costs may vary based on actual capacity and compute needs.

100 GB ⌄	**7**	**30**	**7**
Data Collected per day	Hot Cache retention (days)	Total retention (days)	Estimated Data Compression

◖● AUTO-SELECT ENGINE INSTANCES

⌄ Engine Instances $217.54

⌄ Data Management Instances $83.22

Figure 12.3 – Azure calculator estimates

The Azure pricing calculator does not take into consideration any discounts you may be entitled to if you have an enterprise agreement, but it is still useful for providing cost estimates. The pricing calculator is a tool that I use a lot in my day job to share cost estimates for new environments we build on Azure.

In this section, we learned about the different cluster types and discussed why you should not use the dev/test cluster for production workloads. We then looked at the different VM SKUs and learned that Azure offers different SKUs for different types of workloads. For instance, there are storage-optimized SKUs that are designed for high data ingestion, and there are compute-optimized SKUs for handling large volumes of queries.

In the next section, *Introducing Azure Advisor*, we will discuss an Azure feature that analyzes your Azure usage to help optimize your deployments in terms of cost, security, and operational excellence.

Introducing Azure Advisor

Azure Advisor may be one of the most underrated services on Azure. Azure Advisor provides insights and recommendations, based on your usage, on how to optimize your Azure deployments from a security, cost, performance, and operational excellence perspective. Another great benefit of Azure Advisor is that it is *free*. Simply search for `Azure Advisor` under **All services** in the Azure portal.

As shown in the following screenshot, Azure Advisor provides a clean dashboard showing the number of items that can potentially help optimize your deployment. Each item is ranked based on the impact of the saving:

Figure 12.4 – Azure Advisor

As you can see from the previous screenshot, we have two recommendations under **Cost**, which could save us USD 1,071 annually, and one **Medium impact** recommendation under the **Operational excellence** category. To see more information on the recommendations, simply click on one of the categories to open it. You will then be taken to a detailed list of recommendations. As shown in the following screenshot, one of the recommendations is to reduce the caching policy on our tables. We can get more details on the exact tables by clicking the description:

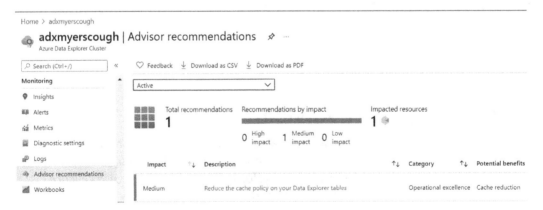

Figure 12.5 – Azure Advisor recommendations for the ADX cluster

As shown in the following screenshot, one of the recommendations is to reduce the caching policy on the `EnglishPremierLeague` table from 30 days to 0, since we have not been querying the data lately:

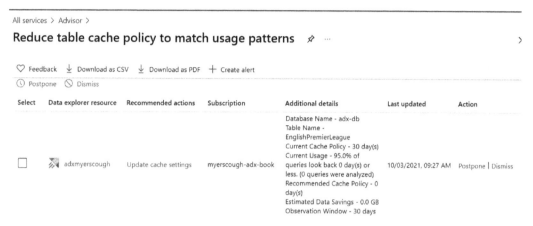

Figure 12.6 – Detailed description of recommendations

As we learned earlier, storing data in the hot cache is expensive, and by making this change, we could see cost savings. As you may recall, we learned how to configure caching policies in *Chapter 11*, *Performance Tuning in Azure Data Explorer*.

In this section, we introduced Azure Advisor and learned that it gives recommendations on how we can optimize our Azure deployments based on our usage. In the next section, *Introducing Cost Management + Billing*, we will discover how to view our invoices, usage, and forecasts, and how to define budgets and alerts.

Introducing Cost Management + Billing

Azure provides some very good cost management services directly in the Azure portal. As mentioned earlier, Azure Advisor analyzes our usage and makes recommendations on how we can save money. As shown in *Figure 12.7*, the **Subscription** overview blade provides a rich view in terms of current spending and forecasts.

The blade also provides the ability to download your invoices:

Figure 12.7 – Subscription billing overview

Let us review each of the tiles on the overview blade, as follows:

1. **Latest billed amount**—This displays the amount of the last invoice. The current invoice can be downloaded using the **Download** button, and a list of historical invoices can be retrieved by clicking **View invoices**.

2. **Invoices over time**—This displays a historical view of the last 4 months' invoices.

3. **Shortcuts**—This provides some shortcut links that allow you to email the invoices and view your resources by spend.

4. **Spending rate and forecast**—This is a very useful graph that displays your current spend and a forecast for the rest of the month.

5. **Costs by resource**—This displays a basic breakdown of cost per resource—for instance, ADX cluster, storage, and load balancers.

The Azure portal also provides a centralized cost management service called **Cost Management + Billing**. From here, you can view your invoices, explore your cost analysis metrics, and create budgets and alerts.

In the next section, we will demonstrate how to view and download your invoices from the **Cost Management + Billing** service.

Accessing invoices

The following sequence of steps demonstrates how to access and download your invoices:

1. Go to `https://portal.azure.com`.

2. Click **All services** and search for `Cost Management + Billing`.

3. Under the properties for **Cost Management + Billing**, click **Cost Management**.

4. As shown in the following screenshot, under the Billing menu in the left-hand pane, click **Invoices**:

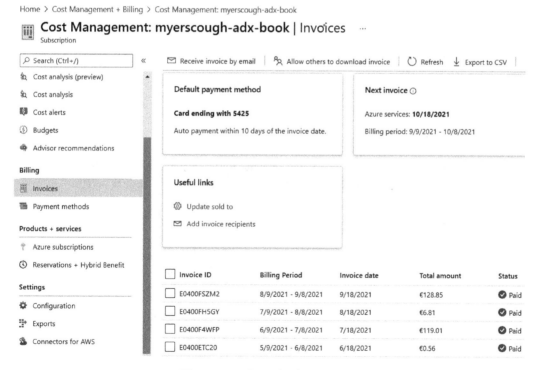

Figure 12.8 – Downloading invoices

5. Select one of your invoices and then click **Download invoices**.

6. Once downloaded, open the **Portable Document Format (PDF)** file. It should look something like this:

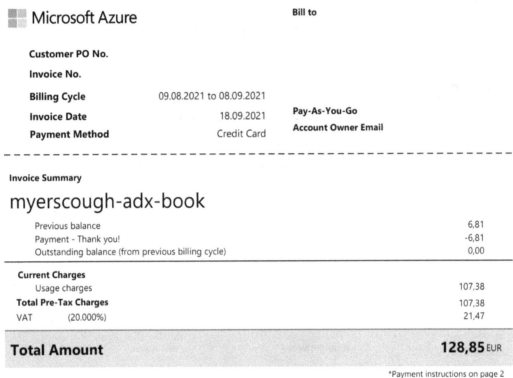

Figure 12.9 – Monthly invoices from Azure

In this section, we learned how to access and download our invoices. In the next section, *Configuring budget alerts*, we will learn how to create budgets and configure alerts.

Configuring budget alerts

The following sequence of steps demonstrates how to create a new budget and configure alert thresholds. In this example, we will be using the action groups we created in *Chapter 9*, *Monitoring and Troubleshooting Azure Data Explorer*:

1. Log in to the Azure portal (`https://portal.azure.com`).

2. Click **All services**, then search for `Cost Management + Billing`.

3. Under properties, click **Cost Management**.

4. As shown in the following screenshot, in the properties pane, click **Budgets** and then click **+ Add**:

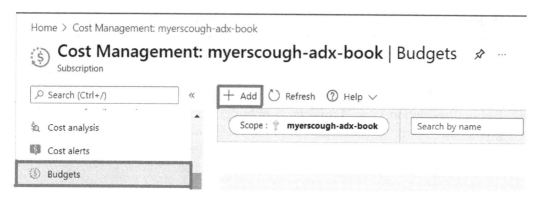

Figure 12.10 – Creating a new budget alert

5.   Let us keep the scope to our subscription—for example, `myerscough-adx-book`—and enter a name for the budget, such as `monthly-subscription-limit`.

6.   Next, under **Budget Amount**, enter an amount. In my case, I have entered `150` EUR, which is my local currency. As shown in the following screenshot, a horizontal dashed line appears on the graph:

Figure 12.11 – Configuring new budget

This indicates our budget that we have just defined.

7.   Click **Next >**. Once the budget has been validated, which can take up to a minute, we can start defining our alert thresholds.

8. Under **Alert conditions**, we define thresholds for our alerts. Once a threshold has been exceeded, an alert will be sent to the action group, which will then trigger an **Azure function** and send emails or **Short Message Service (SMS)** messages. In this example, we will reuse the `AdxAlerts` action group from *Chapter 9, Monitoring and Troubleshooting Azure Data Explorer.*

9. For the **Type** option, select **Actual**; for **% of budget**, enter 50; and for the **Action group** option, select **AdxAlerts**. As shown in the following screenshot, I have set thresholds for 50%, 75%, 80%, 90%, and 95%. An alert will be fired whenever one of these thresholds is exceeded:

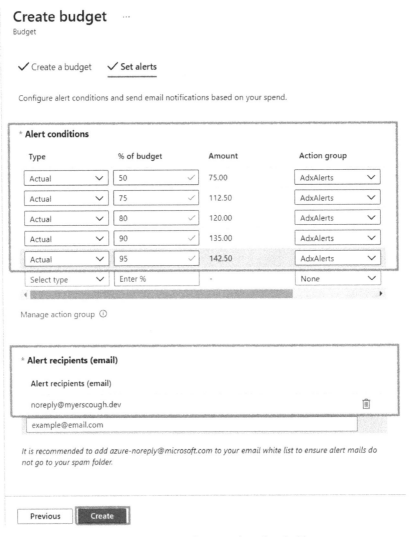

Figure 12.12 – Configuring alert thresholds

The budgets will be overlaid on the **VIEW OF MONTHLY COST DATA** graph, which was shown in *Figure 12.11*.

10. The final step is to add an email address under **Alert recipients (email)**. An email is sent to **Alert recipients (email)** in addition to the event sent to our action group. Click **Create** to complete the budget creation.

> **Note**
>
> Please note that it can take up to 24 hours before a new budget is evaluated, so you may not see your first alert until the following date. After initial creation, budgets are evaluated once every 24 hours.

Once the budgets are in effect, you should start to receive alerts via SMS and email if you used the action group we created in *Chapter 9, Monitoring and Troubleshooting Azure Data Explorer*. Once my alerts came into effect, I started receiving a daily SMS and email informing me that I had exceeded a budget threshold, as shown in the following screenshot:

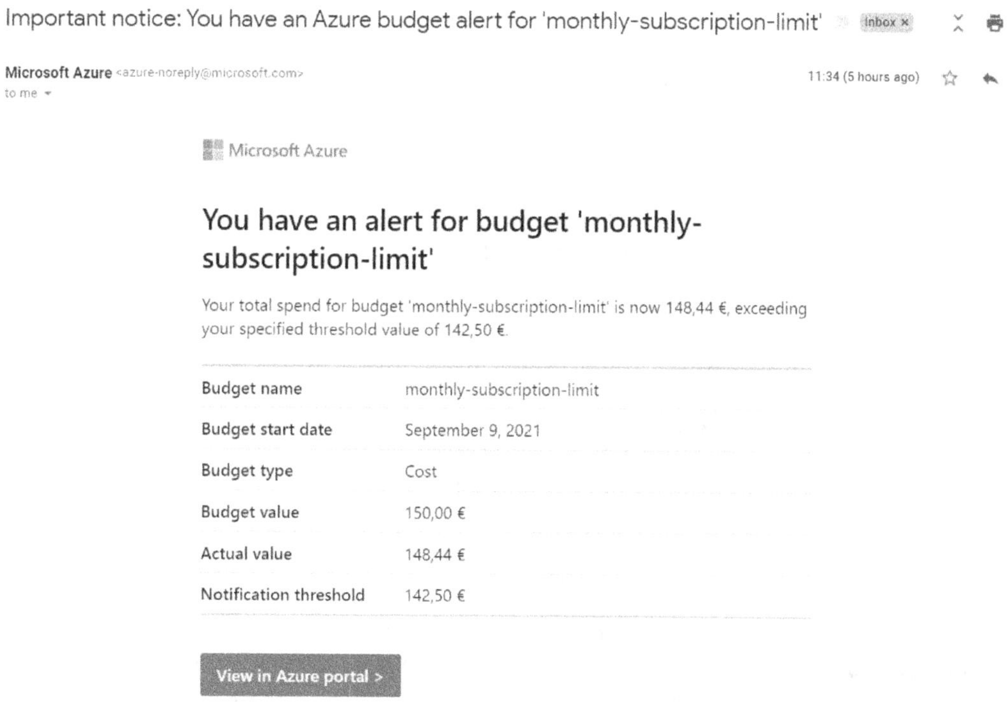

Important notice: You have an Azure budget alert for 'monthly-subscription-limit'   Inbox ×   ✕ 🖨

Microsoft Azure <azure-noreply@microsoft.com>   11:34 (5 hours ago) ☆ ↩
to me ▾

▦ Microsoft Azure

You have an alert for budget 'monthly-subscription-limit'

Your total spend for budget 'monthly-subscription-limit' is now 148,44 €, exceeding your specified threshold value of 142,50 €.

Budget name	monthly-subscription-limit
Budget start date	September 9, 2021
Budget type	Cost
Budget value	150,00 €
Actual value	148,44 €
Notification threshold	142,50 €

`View in Azure portal >`

Figure 12.13 – Email notification for budget alerts

As shown in the following screenshot, I also received an SMS alert each day:

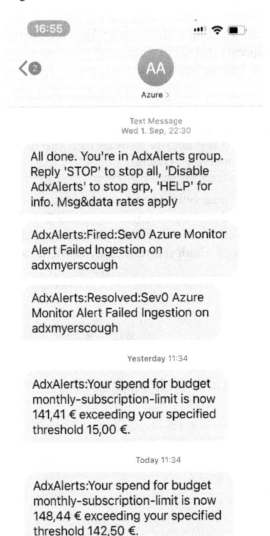

Figure 12.14 – SMS alerts for budgets

In this section, we learned how to configure budgets and alerts to help us manage our spending on Azure. Action group triggers/notifications are a great and simple way to automate responses to triggers by invoking other Azure services such as Azure Functions and Azure Logic Apps. Azure Functions and Azure Logic Apps are beyond the scope of this book, but if you are interested in learning more, then I recommend Packt Publishing's *Understanding Azure Logic Apps* video series and *Learning Azure Functions* book.

Summary

In this chapter, we learned about different ADX cluster SKUs and how to select the correct SKU for your use case. The different SKUs are optimized for data ingestion/storage and high-volume workloads.

We then discussed ADX's scaling capabilities and what you should be aware of when using either manual or automatic scaling, and how workload groups can help reduce the risk of bad-performing queries/jobs impacting how we auto-scale.

Next, we introduced a useful Azure feature called Azure Advisor. Based on our usage patterns, Azure Advisor analyzes our usage patterns and makes cost-saving and performance-related recommendations based on how we use the platform.

Finally, we introduced Azure's cost management and billing features and learned how to view our invoices, generate billing reports and forecasts, and how to configure budgets and alerts so that we are notified when we are close to or have exceeded our budgets.

Before we end the book, I would like to thank you for taking the time to read the contents and try the code examples. I hope you have enjoyed the journey and learned about ADX along the way. This has been a great learning experience for me, and we may cross paths again in the near future.

13
Assessment

Chapter 1

Modify `${Home}/Scalable-Data-Analytics-with-Azure-Data-Explorer/Chapter01/first-query.kql` to display an area chart.

Answer:

The solution can be found in this book's **GitHub** repository:

`${Home}/Scalable-Data-Analytics-with-Azure-Data-Explorer/Chapter01/population-areachart.kql`

Chapter 2

1. Modify `adx-powershell.ps1` and try to deploy another cluster with `doubleEncryption` parameter enabled. Hint: `New-AzKustoCluster` has an optional parameter called `-EnableDoubleEncryption`.

   Answer:

   ```
   New-AzKustoCluster -ResourceGroupName $resource_
   group_name -Name $cluster_name -Location $location
   -SkuName 'Dev(No SLA)_Standard_D11_v2' -SkuTier 'Basic'
   -SkuCapacity 1 -EnableDoubleEncryption
   ```

2. Create a new parameter file and enable purging and `doubleEncryption` parameter, then, change the *hot cache period* to *10 days* and the *soft delete period* to *100 days*.

    Answer:

    The complete solution can be found at the following location: `https://github.com/PacktPublishing/Scalable-Data-Analytics-with-Azure-Data-Explorer/blob/main/Solutions/ch02ex02.params.json`.

    ```
            "adx_cluster_enablePurge": {
              "value": true
            },
            "adx_cluster_enableDoubleEncryption": {
              "value": true
            },
            "adx_db_softDeletePeriod": {
              "value": 100
            },
            "adx_db_hotCachePeriod": {
              "value": 10
            }
    ```

3. In **Azure portal**, create a second **Azure Data Explorer** (**ADX**) database and then set the hot cache period to 10 days and the soft delete period to 50 days.

    Answer: The following screenshot shows how to set these values:

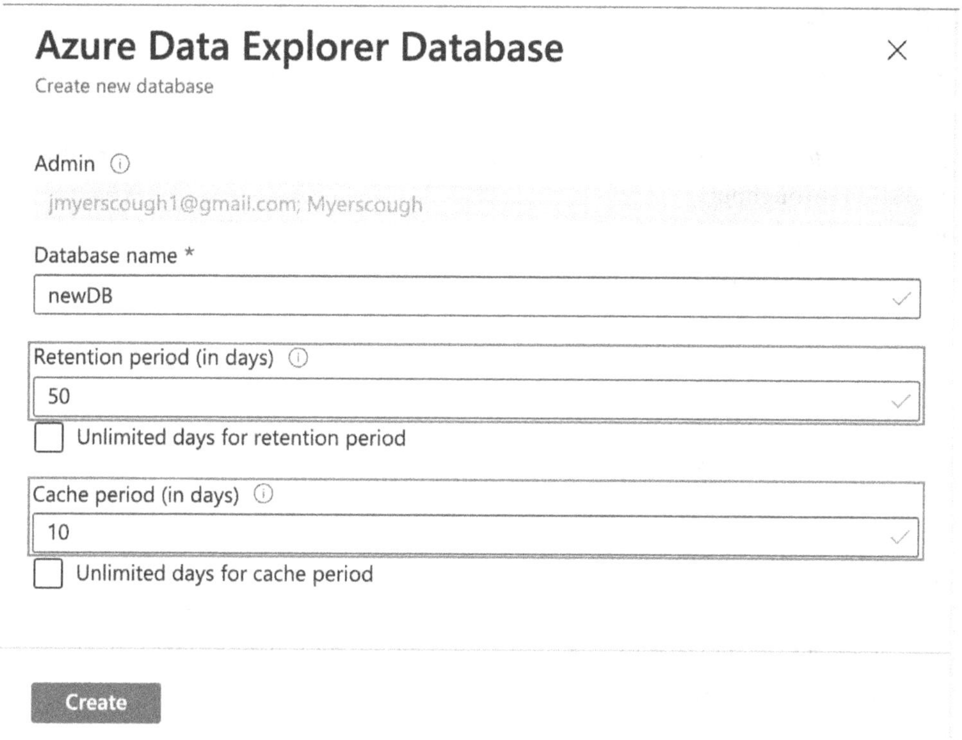

Azure Data Explorer Database   ✕

Create new database

Admin ⓘ

jmyerscough1@gmail.com; Myerscough

Database name *

newDB   ✓

Retention period (in days) ⓘ

50   ✓

☐ Unlimited days for retention period

Cache period (in days) ⓘ

10   ✓

☐ Unlimited days for cache period

Create

Figure 13. 1 – Creating a new ADX database

4. Modify `adx-powershell.ps1` and deploy your ADX cluster to an Azure region that is close to you.

    Answer: `$location = 'eastus'`

5. What is the difference between the hot cache and soft delete retention periods?

    Answer: The hot cache period is the duration for which the data is stored on the actual virtual machine's SSD storage. Soft delete refers to the duration for which the data is stored on the cluster.

6. What shells are supported by **Azure Cloud Shell**?

    Answer: **PowerShell** and **Bash**.

7. How do you open the code editor in Cloud Shell?

    Answer: Type `code   .` in PowerShell.

Chapter 4

1. Which of the following is the preferred ingestion method: *streaming ingestion* or *batch ingestion*?

    Answer: *Batch ingestion* is the preferred ingestion method and is designed for high ingestion throughput.

2. Try to import the `StormEvents` CSV file using the one-click ingestion method. We will use the `StormEvents` table in the next chapter.

    Answer:

    I.   Log in to the ADX Web UI by going to `https://dataexplorer.azure.com/`.

    II.   Click **Add Cluster** and then enter the URL of your ADX cluster. For instance, my ADX instance is called `https://myerscoughadx.westeurope.kusto.windows.net`. Then, click **Add** to connect to the instance.

    III. From the left-hand side menu, click **Data**. Ensure you have expanded your cluster and have clicked your database. This is to ensure you are at the correct scope level, otherwise, the *one-click Ingestion* method will not work.

    IV.   Under **Ingest new data**, click **Ingest**.

    V.   For the **Cluster** option, select your ADX cluster.

    VI.   For the **Database** option, select the **adxdemo** database.

    VII. For the **Table** option, click **Create new**. Then, create a new table called `StormEvents` and click **Next: Source**.

    VIII. For the **Source type** option, select **From file**.

    IX.   Drag your file from your local machine into the web browser. For example, I uploaded the following CSV file: `${HOME}/Scalable-Data-Analytics-with-Azure-Data-Explorer/Chapter04/datasets/stormevents/StormEvents.csv`.

    X.   Click **Next: Schema**.

    XI.   Click **Next: Summary** to begin ingesting the data.

3. Try to upload another Premier League JSON file.

    Answer:

    I.   In the Azure Portal, open the `packtdemo` storage account (or whatever name you gave it earlier if you changed the parameter file).

    II.   Under the **Data Storage** properties, select **Containers** and then click our **results** container.

    III. Select **Upload** and upload one of our JSON files, such as `${HOME}/Scalable-Data-Analytics-with-Azure-Data-Explorer/Chapter04/datasets/premierleague/json/season-1516_json.json`.

4. Update `EPL_Custom_JSON_Mapping` and include the referee.

    Answer: First, we need to update the table:

```
.alter table ['EnglishPremierLeagueJSON'] (Date: string,
HomeTeam: string, AwayTeam: string, FulltimeHomeGoals:
int, FulltimeAwayGoals: int,
    FulltimeResult: int, Referee: string)
```

Now, update the schema:

```
.alter table ['EnglishPremierLeagueJSON'] ingestion json
mapping 'EPL_Custom_JSON_Mapping'
'['
    '{"Column": "Date", "Properties": {"Path":
"$.Date"}},'
    '{"Column": "HomeTeam", "Properties": {"Path":
"$.HomeTeam"}},'
    '{"Column": "AwayTeam", "Properties": {"Path":
"$.AwayTeam"}},'
    '{"Column": "FulltimeHomeGoals", "Properties":
{"Path": "$.FTHG"}},'
    '{"Column": "FulltimeAwayGoals", "Properties":
{"Path": "$.FTAG"}},'
    '{"Column": "FulltimeResult", "Properties": {"Path":
"$.FTR"}},'
    '{"Column": "Referee", "Properties": {"Path":
"$.Referee"}}'
']'
```

Chapter 5

1. Write a query for our `EnglishPremierLeague` data and aggregate the number of matches refereed by each referee.

    Answer:
    ```
    EnglishPremierLeague
     | summarize matches_refereed = count() by Referee
    ```

2. What is the main difference between the `search` and `where` operators?

    Answer: The `search` operator is simple and convenient to use, but you need to be careful with regard to the scope of the search. Searching across multiple tables and columns is an expensive operation and can cause performance issues.

3. Aggregate all the event types in the `StormEvents` table for `California` and render the results as a column chart.

    Answer:
    ```
    StormEvents | where State =~ "California"
     | summarize event=count() by EventType | render
    columnchart
    ```

4. What type of join should you use if you want to include duplicate common column matches in the result set?

    Answer: The `inner` join returns all matches.

Chapter 6

1. What are the properties of a time series?

    Answer:

    - **Trend**: This refers to the long-term direction of the data. For example, the data can have a positive growth known as an *upward trend*, or it can have a negative growth known as a *downward trend*, or the data could also *plateau*.

    - **Variations**: This refers to the peaks and troughs in the data.

    - **Seasonality**: This refers to reoccurring patterns at regular intervals.

    - **Cycles**: These are like seasonality meaning there is a consistent pattern, but the patterns are not consistent at regular time intervals.

2. What operator can we use to generate a time series?

    Answer: The `make-series` operator.

3. Can you fill in the blanks of this query to display the number of patches installed in the last 30 days and render the results as a time chart?

```
let startTime = ago(____);
let endTime = now();
let binSize = 7d;
Update
| where Classification == "Security Updates"
| make-series security_updates=count() default=0 on
TimeGenerated from startTime to endTime step _____ by
UpdateState
| render ____
```

Answer:

```
let startTime = ago(30d);
let endTime = now();
let binSize = 7d;
Update
| where Classification == "Security Updates"
| make-series security_updates=count() default=0 on
TimeGenerated from startTime to endTime step binSize by
UpdateState
| render timechart
```

4. Using mv-expand, split the following time series into records:

```
let startTime = ago(100d);
let endTime = now();
let binSize = 7d;
Update
| where Classification == "Security Updates"
| make-series security_updates=count() default=0 on
TimeGenerated from startTime to endTime step binSize by
UpdateState
```

Answer:

```
let startTime = ago(100d);
let endTime = now();
let binSize = 7d;
```

```
Update
| where Classification == "Security Updates"
| make-series security_updates=count() default=0 on
TimeGenerated from startTime to endTime step binSize by
UpdateState
| mv-expand TimeGenerated to typeof(datetime), security_
updates to typeof(long)
| project UpdateState, security_updates, TimeGenerated
| order by UpdateState
```

Chapter 7

1. What is the purpose of *moving averages*?

    Answer: *Moving averages* allow us to remove noise and smooth our data.

2. What is the purpose of *linear regression*?

    Answer: The purpose of *linear regression* is to identify trends – both positive and negative – in our time series.

3. What are the extra steps required to render time charts in log analytics?

    Answer: With log analytics, we must expand our time series using mv-expand and convert our values from dynamic data types to their specific data types. Then, we must project the columns we want to pipe to the render operator.

4. In *Figure 7.12*, we rendered an anomaly chart to display the anomalies in the time series. Using series_fir(), generate a smoother graph without the anomalies. Once you generate a smoother output, pass your data to series_decompose_anomalies() to see if there are still any anomalies. The query to generate the graph in *Figure 7.12* is as follows. You will need to connect to the help cluster (https://help.kusto.windows.net/) to complete this exercise:

```
let startTime = toscalar(demo_make_series1 | summarize
min(TimeStamp));
let endTime = toscalar(demo_make_series1 | summarize
max(TimeStamp));
let binSize = 1h;
demo_make_series1
```

```
| make-series requests=count() default=0 on TimeStamp
from startTime to endTime step binSize
  | extend anomalies = series_decompose_anomalies(requests)
  | render anomalychart with (anomalycolumns=anomalies)
```

Answer:

```
let startTime = toscalar(demo_make_series1 | summarize
min(TimeStamp));
let endTime = toscalar(demo_make_series1 | summarize
max(TimeStamp));
let binSize = 1h;
let windowSize = 20;
demo_make_series1
  | where Country == "United States"
  | make-series requests=count() default=0 on TimeStamp
from startTime to endTime step binSize by Country
  | extend movingAverage = series_moving_avg_fl(requests,
windowSize, true)
  | extend anomalies = series_decompose_
anomalies(movingAverage)
  | render anomalychart with (anomalycolumns=anomalies)
```

Chapter 8

1. What is *data visualization*?

   Answer: *Data visualization* is a method of illustrating data. This can help different audiences understand your data.

2. Create a dashboard in ADX for the EnglishPremierLeague table and display the total number of goals scored, goals conceded, and wins for each instance You can experiment here and add some extra tiles.

Answer:

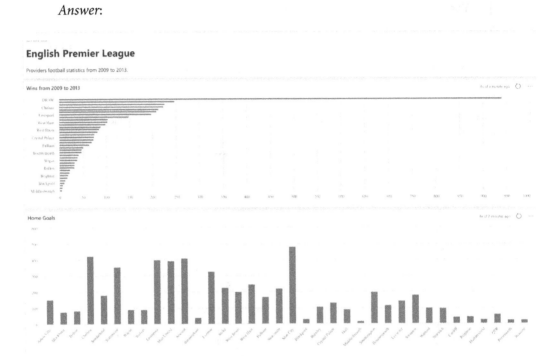

Figure 13.2 – A chart displaying new tiles for the EnglishPremierLeague table

3. What is the purpose of *parameters* in ADX dashboards?

    Answer: Parameters allow the end user to interact with your dashboard and change filter values.

4. What is the difference between the **Import** and **DirectQuery** functions when configuring the ADX connector for **Microsoft Power BI**?

    Answer: The **Import** function imports your data into Power BI, whereas the **DirectQuery** function pulls data directly from the ADX cluster.

5. Try modifying the dashboard we created in ADX Web UI and update the title by editing the markdown text.

    Answer:

    I. Click **Edit** and then click the text field's pencil icon in the far right to enter edit mode.

    II. As per *Figure 8.18*, edit the text field.

    III. Enter the following markdown:

    ```
    # Storm Summary for the United States
    _The following dashboard provides a summary of the
    direct injuries and deaths, total property damage, an
    aggregation of all the storm events in the United States
    and the location of each storm._
    ```

    IV. Click **Apply changes** and then click **Save changes** to save the dashboard.

Chapter 9

1. What is the difference between SLIs, SLOs, and SLAs?

    Answer:

    - **Service level indicators**: SLIs are the individual indicators that you are monitoring. An SLO could depend on multiple SLIs. In the context of ADX, a few examples of SLIs are CPU usage, ingestion latency, and hot cache size.

    - **Service level objectives**: SLOs are (typically internal) goals or objectives for a team – for example, ensuring the response time is less than 5 milliseconds. The SLOs are essentially the requirements for your alerting thresholds and conditions.

    - **Service level agreements**: SLAs are the agreements you make with your customers and you would typically put them into contracts. For example, ensuring the uptime of your system is a common SLA.

2. Configure a metrics dashboard to display the Blobs Received metric and then import some data into your ADX cluster. What do you see?

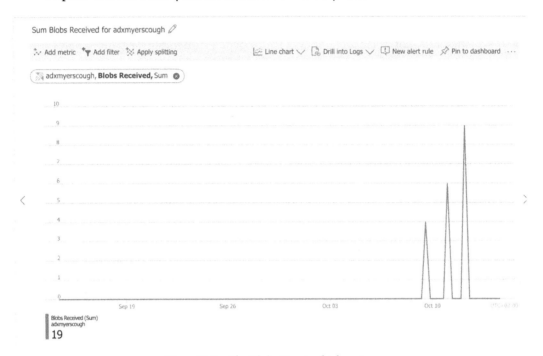

Figure13.3 – The Blobs Received telemetry

3. Try to implement monitoring alerts for the event hub's incoming messages and set their severity to **Informational**, as this instance is typically not an error.

    Answer:

    I.   Log in to the Azure portal (`https://portal.azure.com`).

    II.   Click on your Event Hubs namespace – for instance, `adxeventhubnamespacemyerscough`.

    III. Under the Event Hubs namespace's properties, click **Metrics**.

    IV.   From the **Metrics** drop-down menu, select **Incoming Requests**.

    V.   Under the **Condition name**, click **Whenever the sum incoming requests is greater than <logic undefined>**.

    VI.   Set the value under **Threshold value** to 1 and click **Done**.

    VII. Click **Manage action groups** to assign an Action Group. For instance, I reused `AdxAlerts`.

VIII. Give the alert a name and a description.

IX.   Set the **Severity** option to **3 – Informational**.

X.   Click **Create alert Rule**.

4. How many severity levels are there and what does each level mean?

Answer:

Level	Severity
0	Critical
1	Error
2	Warning
3	Informational
4	Verbose

Chapter 10

1. Assign the `contributor` role to one of your **Azure Active Directory** (**AAD**) users in the management plane.

Answer:

I.   Open the Azure portal (`https://portal.azure.com`) and click the subscriptions icon.

II.   If you do not have the subscriptions icon pinned to your sidebar, click **All services** and search for `subscriptions`.

III.   Click **Subscriptions** and then click your subscription (for example, **myerscough-adx-book**).

IV.   In the properties pane, click **Access control (IAM)**.

V.   Click **+ Add** then click **Add role assignment**.

VI.   On the **Add role assignment** blade, select the **Contributor** role from the **Role** drop-down menu.

VII.   Select **Assign access to** to be **User, group, or service principal**.

VIII. In the **Select** textbox, search for your user (for example, `Sofia`) and click on the user record.

IX.   Click **Save** to add the user.

2. Assign the `Database ingestor` role to a user.

    Answer:

    I.   Go to `https://portal.azure.com` and click on your ADX cluster
          (for example, you might click on **adxMyerscough**).

    II.   Under your cluster's properties, click **Databases**.

    III. Click the database you would like to add a user to (for example, you might click
          on **adx-db**).

    IV.   Under your database's properties, click **Permissions**.

    V.   Click + **Add**, and from the drop-down menu, select **Ingestor**.

    VI.   Search for a user in your Azure Active Directory. For example, I am going
          to make a user called `Lukas` one of the database ingestors.

    VII. Click **Select** once you have selected your user to complete the process. Now,
          you should see the user listed as an administrator.

3. What happens when you log in to the ADX Web UI and try to query the database
    as that user?

    Answer: The user is unable to query the databases.

Chapter 11

1. What is the purpose of workload groups?

    Answer: Workload groups allow us to group together requests and queries and allow
    us to define constraints such as request limits, connection limits, and resource limits
    via workload group *policies*. This allows us to control access, monitor performance,
    and prevent users from consuming all of our clusters' resources.

2. Assuming we have our request classification policy configured and enabled, what
    will happen when we execute the following query as a database administrator?

    ```
    .alter cluster policy request_classification
    '{"IsEnabled":false}' <|
        iff(current_principal_is_member_
    of('aadgroup=TrialUsers;27447925-1f0e-41b6-b01f-
    973eaab478b0'), "Packt Demo","default")
    ```

    Answer: The request classification policy will be disabled and all requests will be
    assigned to the default workload group.

3. Why should you filter your data based on a date field as early as possible in your query?

    Answer: If your datasets contain a date field, ensure your first where clause filters the time range you are interested in. This greatly improves your query performance, and the improvements are related to how ADX stores data. In *Chapter 1*, *Introducing Azure Data Explorer*, we learned that ADX stores our data in *shards*, also known as *extents*. By filtering our data with a specific time range, we exclude data outside of the time range, thereby reducing the amount of data processing during execution time.

4. Create a dashboard in the ADX Web UI and display the query execution metrics, such as the top five longest running queries, then, aggregate the workload groups.

    Hint: Use .show queries and review *Chapter 8*, *Data Visualization with Azure Data Explorer and Power BI*.

    Answer:

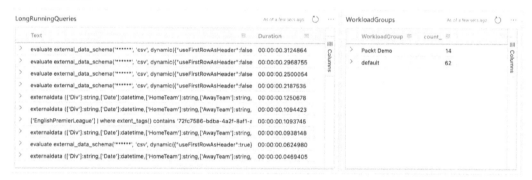

Figure 13.3 – Displaying query execution metrics

Index

U

V

W

Packt.com

Subscribe to our online digital library for full access to over 7,000 books and videos, as well as industry leading tools to help you plan your personal development and advance your career. For more information, please visit our website.

Why subscribe?

- Spend less time learning and more time coding with practical eBooks and Videos from over 4,000 industry professionals

- Improve your learning with Skill Plans built especially for you

- Get a free eBook or video every month

- Fully searchable for easy access to vital information

- Copy and paste, print, and bookmark content

Did you know that Packt offers eBook versions of every book published, with PDF and ePub files available? You can upgrade to the eBook version at packt.com and as a print book customer, you are entitled to a discount on the eBook copy. Get in touch with us at customercare@packtpub.com for more details.

At www.packt.com, you can also read a collection of free technical articles, sign up for a range of free newsletters, and receive exclusive discounts and offers on Packt books and eBooks.

Other Books You May Enjoy

If you enjoyed this book, you may be interested in these other books by Packt:

Limitless Analytics with Azure Synapse

Prashant Kumar Mishra

ISBN: 978-1-80020-565-9

- Explore the necessary considerations for data ingestion and orchestration while building analytical pipelines
- Understand pipelines and activities in Synapse pipelines and use them to construct end-to-end data-driven workflows
- Query data using various coding languages on Azure Synapse
- Focus on Synapse SQL and Synapse Spark
- Manage and monitor resource utilization and query activity in Azure Synapse
- Connect Power BI workspaces with Azure Synapse and create or modify reports directly from Synapse Studio
- Create and manage IP firewall rules in Azure Synapse

Cloud Scale Analytics with Azure Data Services

Patrik Borosch

ISBN: 978-1-80056-293-6

- Implement data governance with Azure services
- Use integrated monitoring in the Azure Portal and integrate Azure Data Lake Storage into the Azure Monitor
- Explore the serverless feature for ad-hoc data discovery, logical data warehousing, and data wrangling
- Implement networking with Synapse Analytics and Spark pools
- Create and run Spark jobs with Databricks clusters
- Implement streaming using Azure Functions, a serverless runtime environment on Azure
- Explore the predefined ML services in Azure and use them in your app

Packt is searching for authors like you

If you're interested in becoming an author for Packt, please visit `authors.packtpub.com` and apply today. We have worked with thousands of developers and tech professionals, just like you, to help them share their insight with the global tech community. You can make a general application, apply for a specific hot topic that we are recruiting an author for, or submit your own idea.

Share Your Thoughts

Now you've finished *Scalable Data Analytics with Azure Data Explorer*, we'd love to hear your thoughts! Scan the QR code below to go straight to the Amazon review page for this book and share your feedback or leave a review on the site that you purchased it from.

`https://packt.link/r/1-801-07854-8`

Your review is important to us and the tech community and will help us make sure we're delivering excellent quality content.